Midjourney AI绘画教程

设计与关键词创作技巧588例

雷波◎著

·北京·

内 容 简 介

本书讲解了以下三部分内容。第一部分讲解了 Midjourney 的提示词语法结构、撰写逻辑与所有参数、命令的使用方法。第二部分讲解了 Photoshop 新增的 AI 生成功能，并示范了如何组合使用 Midjourney 与 Photoshop 以得到更完美的图像。第三部分为笔者针对 40 个创作领域，给出的 588 个有创作提示语的示例。值得一提的是，本书讲解了 Midjourney 最新的局部重绘、无限扩图与平移功能，此外还讲解了包括换脸在内的两种获得同一角色的技术。

本书旨在满足广泛的读者需求，从 Midjourney 新手到数字艺术家、设计师、摄影师、插画师、广告媒体人，以及教育者和学生，都能受益。它不仅具有实用价值，还具有教材属性，为人工智能和视觉设计领域提供了全面的知识解决方案。

本书还附赠了三本独家原创的中英对照 Midjourney 创作电子书，极大地拓展了本书的知识容量。

图书在版编目（CIP）数据

Midjourney AI 绘画教程：设计与关键词创作技巧 588 例 / 雷波著 .—北京：化学工业出版社，2024.2

ISBN 978-7-122-44444-8

Ⅰ . ① M… Ⅱ . ①雷… Ⅲ . ①图像处理软件－教材 Ⅳ . ① TP391.413

中国国家版本馆 CIP 数据核字（2023）第 217107 号

责任编辑：李　辰　孙　炜　　　封面设计：异一设计
责任校对：刘曦阳　　　装帧设计：盟诺文化

出版发行：化学工业出版社（北京市东城区青年湖南街13号　邮政编码100011）
印　　装：北京瑞禾彩色印刷有限公司
710mm×1000mm　1/16　印张 13　字数 308 千字　2024 年 2 月北京第 1 版第 1 次印刷

购书咨询：010-64518888　　　售后服务：010-64518899
网　　址：http://www.cip.com.cn
凡购买本书，如有缺损质量问题，本社销售中心负责调换。

定　　价：88.00元

前　言

时至今日，Midjourney 仍然是功能最强、出图效果最好的 AI 绘图平台之一，虽然其技术相对并不算太复杂，但初学者仍然会碰到许多问题，例如：

如何让 Midjourney 理解创作者的提示语?

如何获得同一角色的不同角度图像?

Midjourney 不同版本的模型间有什么差异?

如何只修改图像的局部?

如何获得特定的内容?

如何让每次生成的四格图像区别更大?

……

实际上，这些问题笔者在学习道路上都曾经遇到过，但经过大量实践均找到了可供初学者借鉴的经验，本书正是这些经验的汇总，笔者将其分为三个部分。

本书的第一部分是第 1 章与第 2 章，这两章详细讲解了 Midjourney 的参数、命令、提示词语法及撰写逻辑，并辅以大量对比性示范，以便初学者快速入门。

值得一提的是，笔者不仅讲解了 Midjourney 最新的 Niji 模型的四个场景参数，还讲解了 Midjourney 最新的超强局部重绘、无限扩图与平移功能。除此之外，针对 Midjourney 默认无法生成同一角色的难题，提供了包括换脸在内的两种解决方法。

本书的第二部分是第 3 章，笔者详细讲解了 Photoshop 新增的 AI 生成功能，并示范了如何将其与 Midjourney 组合起来使用，以得到更完美的图像。

本书余下的章节为第三部分，在这一部分中，笔者在摄影素材创作、珠宝设计、文创设计、包装设计、箱包设计、珠宝设计等 40 个领域，示范了有完整提示语的总共 588 个不同效果图像的创作范例，这也是本书的最大价值所在。

通过复制这些提示词，然后更换其中的关键词，就可以快速创作出符合自己想法及特定需求的图像，节省了大量摸索、尝试的时间。

书中所有图像均由 Midjourney 渲染生成，几乎每张展示图都配有提示语，并在关键词下方有下划线，以确保读者在使用此关键词时能够得到类似的效果。

为了扩展本书内容，笔者将赠送 3 本拥有海量 Midjourney 常用关键词的 PDF 电子书，下载方法可参考封底。

在人工智能技术飞速发展的今天，想要在这个领域保持竞争力，获得最新、最前沿的技术信息，各位读者必须对新技术保持敏感度与好奇心。可以添加本书交流微信 hjysysp 与笔者团队在线沟通交流，搜索并关注笔者的微信公众号“好机友摄影”，或在今日头条或百度、抖音、视频号中搜索并关注“好机友摄影”或“北极光摄影”。

著　者

目　录

CONTENTS

第1章　掌握Midjourney工作流程及常用参数

第 2 章 掌握 Midjourney 提示语撰写逻辑及常用命令

第 3 章 用 Photoshop AI 功能对 Midjourney 图像局部精调

第 4 章 Midjourney 写实摄影照片创作 76 例

第 5 章 Midjourney 抽象纹理设计 28 例

第 6 章 Midjourney 箱包设计 17 例

第 7 章 Midjourney 服装鞋子设计 29 例

第 8 章 Midjourney 不同风格插画绘制 60 例

第 9 章 Midjourney 珠宝设计 75 例

第 10 章 Midjourney 机甲、铠甲与生肖设计 52 例

第 11 章 Midjourney 特效文字设计效果 48 例

第 12 章 Midjourney 室内设计 26 例

第 13 章 Midjourney 游乐场、卖场、建筑外观创意 27 例

第 14 章 Midjourney 艺术家风格 76 例

第 15 章 Midjourney 版式、盲盒、包装、UI 等设计 74 例

第 1 章　掌握 Midjourney 工作流程及常用参数

掌握 Midjourney 命令区的使用方法

Midjourney（以下正文简称 MJ）生成图像的操作是基于命令或带参数的命令来实现的，当进入 Discord 界面后，在最下方可以看到命令输入区域，在此区域输入英文符号 /，则可以显示若干个命令，如下图所示。

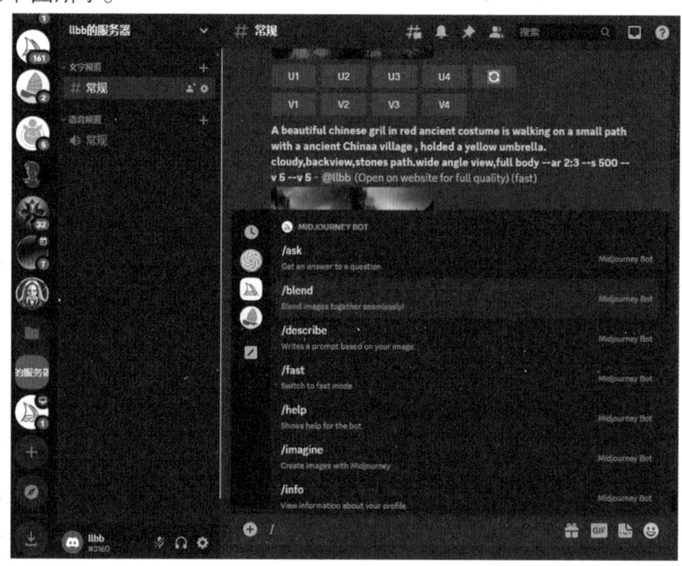

可以在此直接选择某一个命令执行，也可以直接在 / 符号后输入拼写正确的命令。如果被选中的命令需要填写参数，则此命令后面会显示参数类型，如左下图所示的 /blend 命令，如果命令可以直接运行无须参数，则命令显示如右下图所示。

需要注意的是，前面提到的“参数”是一个广义词，根据不同的命令，参数可能是一段文字，也可能是一张或多张图像。

在实际应用过程中，可以通过在 / 符号后面输入命令首字母或缩写的方法，快速显示要使用的命令，例如，对于使用频率最高的 /imagine，只需要输入 /im，就能快速显示此命令，如左下图所示。

如果点击命令行左侧的 + 符号，可以显示如右下图所示的菜单，使用其中的三个命令，可以完成上传图像、创建子区及输入 / 符号等操作。

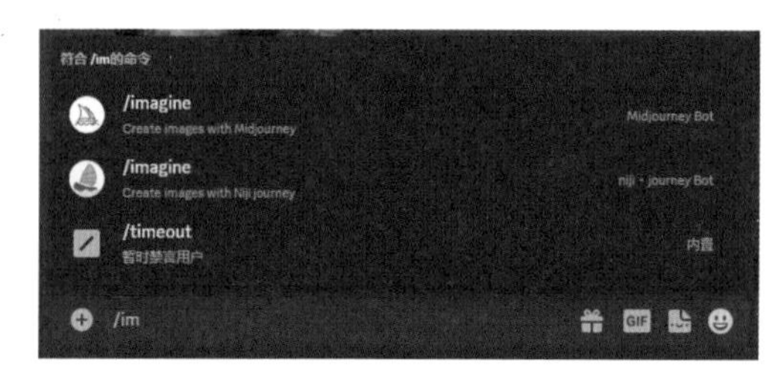

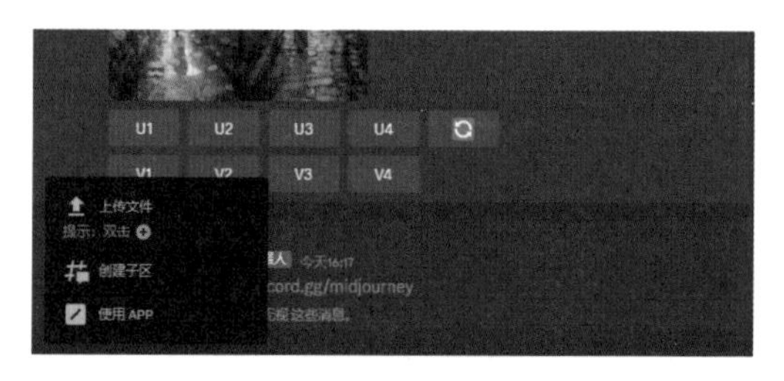

用 imagine 命令生成图像

基本用法及提示语结构

/imagine 命令是 MJ 中最重要的命令，在 MJ 的命令提示行中找到或输入此命令后，在其后输入提示语，即可得到所需的图像，如下图所示。

在 /imagine 命令后面英文部分 chinese dragon with gold helmet, rushing with a scared face. towards the camera frantically. photorealistic . 用于描述要生成的图像。

后面的 --s 1000 --q 2 --ar 16:9 --v 5 是参数，会影响图像画幅、质量和风格等方面。

使用此命令会生成 4 张图像，如右图所示，这 4 张图像被称为四格初始图像，点击后可以放大观看细节。

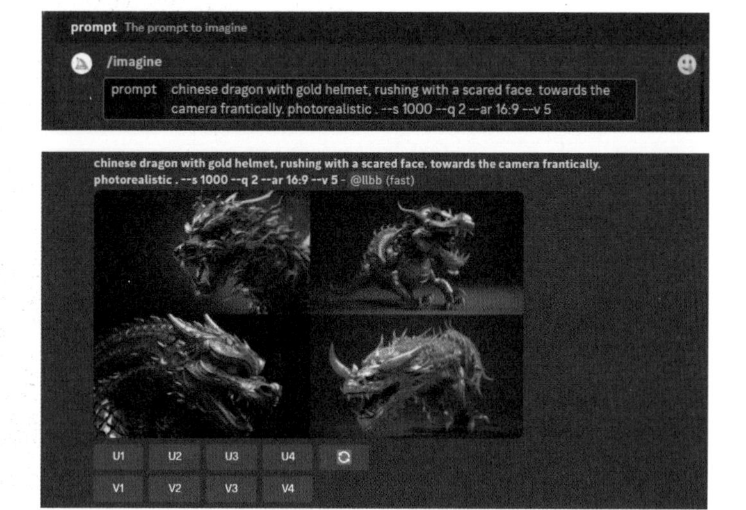

U 按钮与 V 按钮的使用方法

如果认为初始图像效果不错，可以单击 U1 ~ U4 按钮，放大各个初始图像，以得到高分辨率图像。

U1 对应的是左上角图像、U2 对应的是右上角图像、U3 对应的是左下角图像、U4 对应的是左下角图像。如果对于初始图像不太满意，可以单击 V1 ~ V4 按钮，对各个初始图像做衍生操作，使 MJ 针对此初始图像做变化操作。

例如，笔者点击 V1 后，会得到如右图所示的衍变四格图像，在此基础上还可以再分别多次单击 V1 ~ V4 按钮，使 MJ 针对四格图像做再次衍变处理。

如果所有四格图像无法令人满意，可以单击刷新按钮，生成新的四格图像。

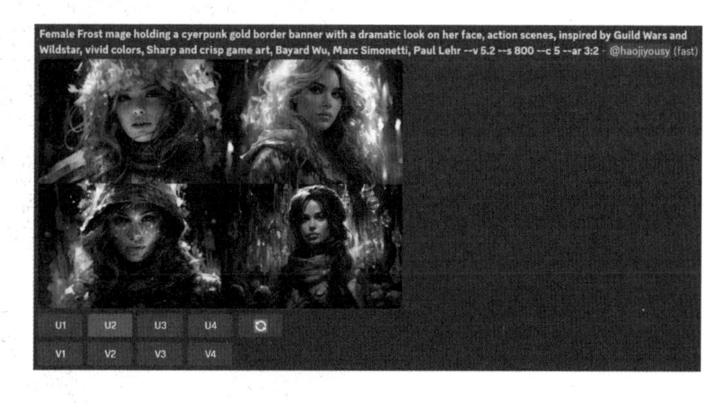

当在四格图像中找到最终满意的图像后，可以单击 U1 ~ U4 按钮，生成高分辨率图像。

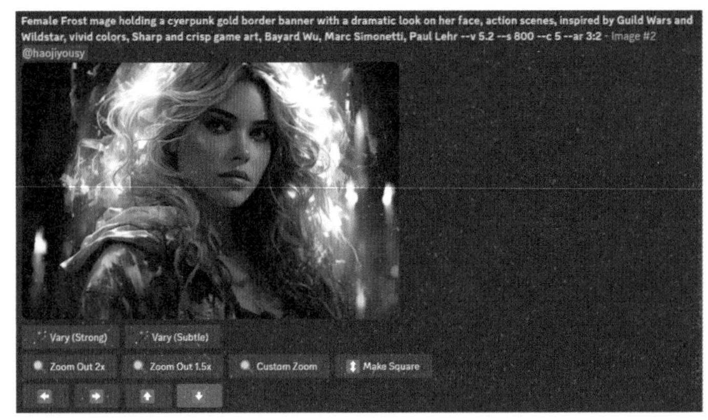

再次衍变操作

要在大图基础上再次执行衍变操作，可以单击 Vary(Strong) 以产生变化幅度更大的四格图像，或单击 Vary(Subtle) 以产生变化更微妙的四格图像。

查看详情操作

如果想要查看此图像的详情，可以单击大分辨率图像下方的 Web 按钮。

此时会进入作品查看页面，在此页面中不仅可以看到提示词、参数和图像分辨率，还能看到许多同类图像，可以通过查看优秀同类图像的提示词，来修正自己的提示词，以得到更优质的图像。

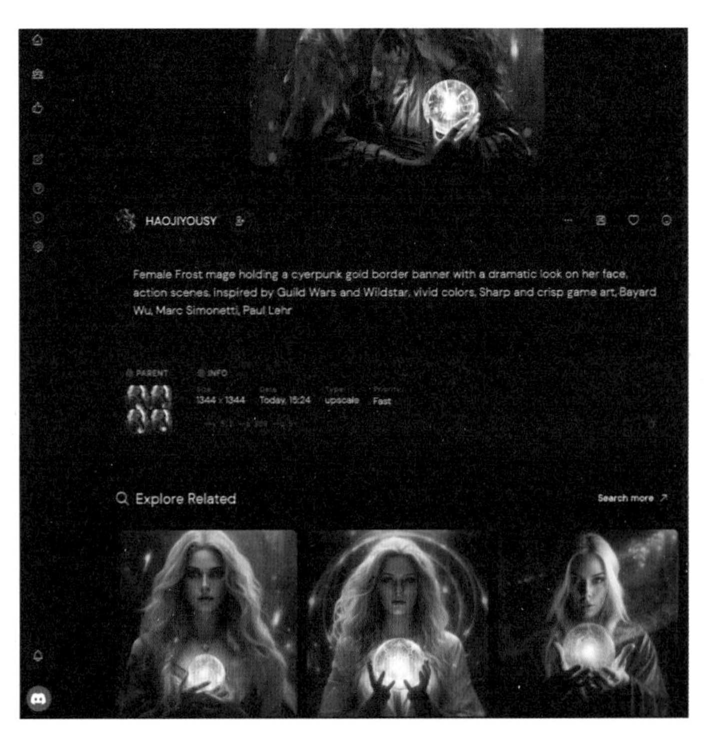

Zoom Out 按钮的使用方法

MJ 更新的 5.2 版提供了强大的 Zoom out 功能，使用此功能可以无限扩展原始图像，这个功能类似于目前许多 AI 软件提供的扩展画布功能。

例如，左下方为原图，在此基础上，可以连续扩展为下面展示的一系列图像，从而使要表现的场景不断扩大。

这意味着，对于初级 MJ 创作者来说，在撰写提示词时不必过于纠结关于景别的单词是否描述正确，只要获得局部图像，就通过使用此功能得到全景图像。

但对于高级创作者来说，必须清晰的是，使用这种方法获得的全景图像与使用正确的全景景别提示词，所获得的图像在透视效果上还有较大的区别。

此功能的方法是先按常规方法获得四格初始图像，如左下图所示，点击 U 按钮生成大图。

然后点击图像下方的 Zoom Out 2x 或 Zoom Out 1.5x 按钮，如右下图所示。如果希望获得其他的放大倍率，点击 Custom Zoom 按钮，并在 --zoom 后面填写 1.0~2.0 之间的数值。

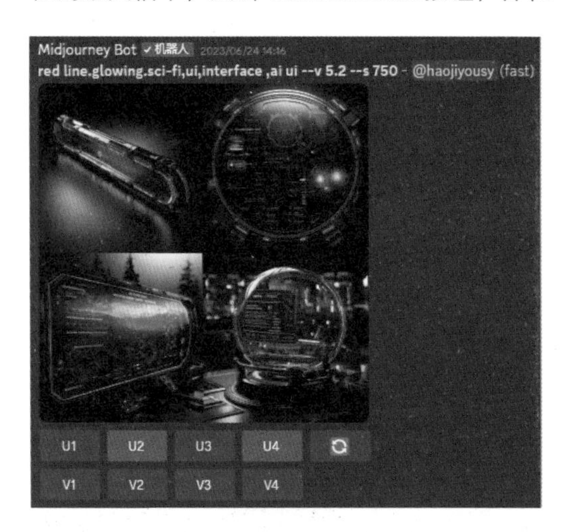

Pan 按钮的使用方法

Pan 按钮是指在 MJ 放大的图像下方的四个箭头图标 ，如左下图所示。其作用类似于前面讲解的 Zoom Out 按钮，用于向某一个方向扩展画面。

这一个功能弥补了 Zoom Out 按钮只能向四周四扩展画面不足，使画面扩展更加灵活，例如，右下图为笔者点击向右箭头扩展画面得到的效果。

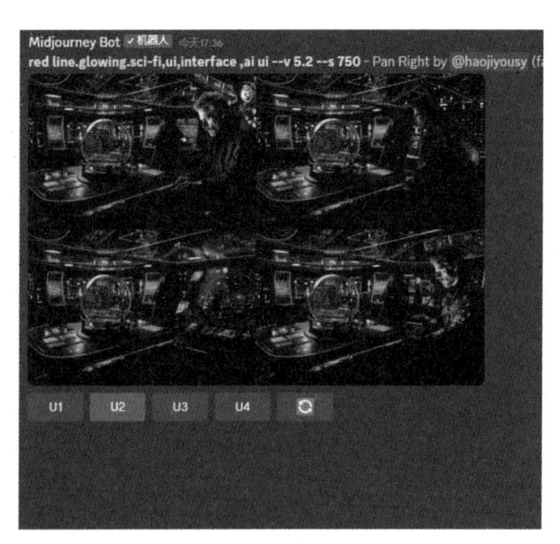

针对扩展得到的四格新图像，可以在点击 U 按钮放大后，再次进行扩展。

但需要注意的是，目前 MJ 不支持对同一图像同时在垂直及水平方向上扩展，因此，点击向右箭头后，在生成的大图下方只能看到两个水平方向的按钮，如左下图所示。

此时创作者可以选择点击 Make Square 按钮将此图像扩展为正方形图像，或继续水平扩展图像，得到如右下图所示的效果。

按此方法经过多次操作，可以获得类似于全景照片式的更大分辨率图像，或用这些素材生成类似于平移镜头扫视的视频。

保存高分辨图像操作

可以通过两种方法来保存最终生成的高分辨率图像。

第一种方法是单击最终生成的高分辨率图像，然后单击左下角的“在浏览器中打开”链接，再单击鼠标右键，在弹出的快捷菜单中选择“图像另存为”命令。

第二种方法是在查看详情页面中单击🖫按钮。

V4 版本与 V5 版本衍变操作的区别

目前在 MJ 中使用最多的是 V4 与 V5 版本，但这两个版本在生成四格初始图像及衍变图像时，区别很大。

使用 V5 版本生成四格初始图像时，所有图像均为高分率图像，因此，单击 U1 ~ U4 按钮后，MJ 只是对四格初始图像进行了切分及轻微的放大操作，因此订阅的时间会更快。

但使用 V4 版本生成的四格初始图像分辨率及质量较低，因此，如果要得到高分辨率图像，必须使用 U1 ~ U4 按钮进行放大操作。

右下图所示为笔者单击右上图下方的 U2 按钮后，生成的图像效果。

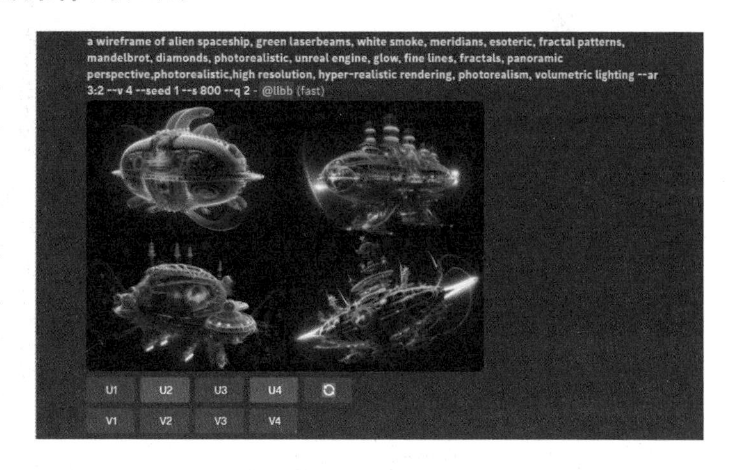

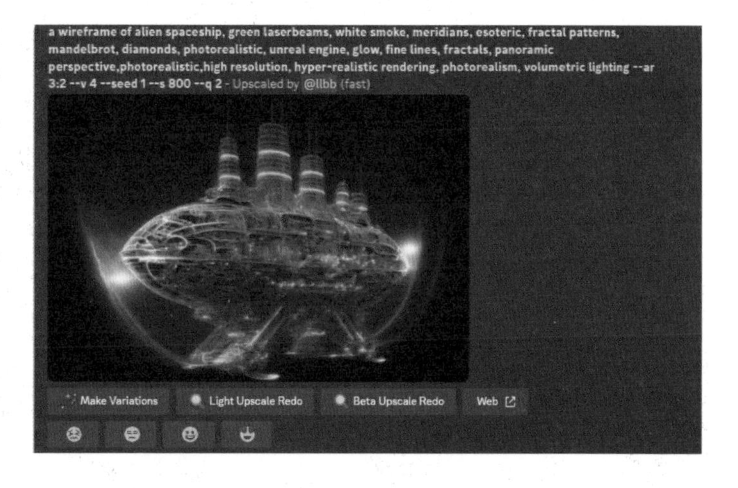

V4 版本 3 种放大操作的区别

如前所述，使用 V4 版本生成的四格初始图像分辨率及质量较低，因此，如果要最终生成高分辨率的图像，必须使用 U1 ～ U4 按钮进行放大操作。

此时，可以选择 3 种放大模式。

其中单击 U1 ～ U4 按钮执行的是默认模式放大，生成的高分辨率图像的宽度最大尺寸为 1024 像素。

此时，如果单击放大后图像下方的 Light Upscaler Redo 按钮，则可以再次放大此图像，只在放大时，MJ 会平滑图像，如果生成的人脸较粗糙，或图像中有光滑的对象，可以考虑使用这种方法，不过根据测试，当有人的全脸时，使用这种方法会轻微破坏人脸结构。

如果单击放大后图像下方的 Beta Upscaler Redo 按钮，则可以执行第 3 种放大图像的操作模式，可将图像的宽度像素放大至 2048 像素，这种模式可增加图像肌理。

下面最左图为 V4 版本下生成的初始图像。

左二图为单击 U 按钮按默认模式放大的图像效果。

右二图为单击 Light Upscaler Redo 按钮得到的效果，可以看到口唇处有明显失真，但皮肤能够得到平滑处理。

最右图为单击 Beta Upscaler Redo 按钮得到的效果图，可以看到口唇处也有失真，且皮肤与衣服处被增加大量肌理。

掌握 Midjourney 生成图像的参数

理解参数的重要性

如前所述，在使用 MJ 生成图像时，需要使用参数控制图像的画幅、质量、风格，以及用于生成图像的 MJ 版本。正确运用这些参数，对于提高生成图像的质量非常重要。

例如，左下图与右下图使用的提示语与大部分参数均相同，只是左下图使用了 --v 5 参数，右下图使用了 --niji 5 参数，从而使得到的两组图像风格截然不同。

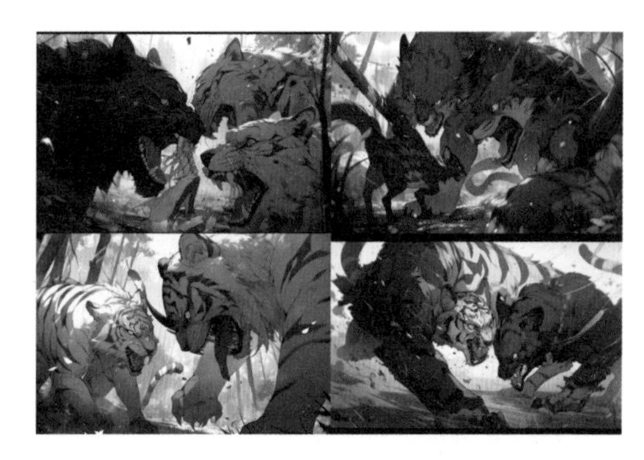

参数撰写方式

在提示语后面添加参数时必须使用英文符号，而且要注意空格问题。

例如，--iw 0.5，不能写成为 --iw0.5，否则 MJ 就会报错。在右侧所示的两个错误消息中，MJ 提示 --v5 与 --s800 格式有误，应该为 --v 5 与 --s 800。

另外，参数的范围也要填写正确，例如，在右侧所示的错误中，MJ 提示在 V5 版本中 --iw 的数值范围为 0.5 ~ 2，因此填写 0.25 数值是错误的。各参数的范围在后面的章节中均有讲解。

随着 MJ 的功能逐渐完善、强大，还会有更多新的参数，但只要学会观看 MJ 的错误提示信息，就能轻松地修改参数填写错误。

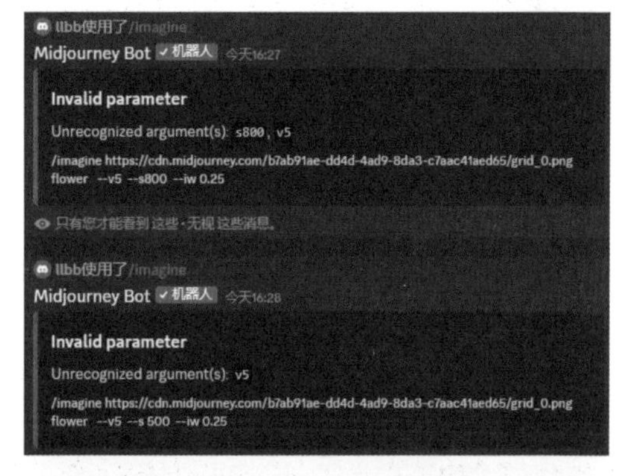

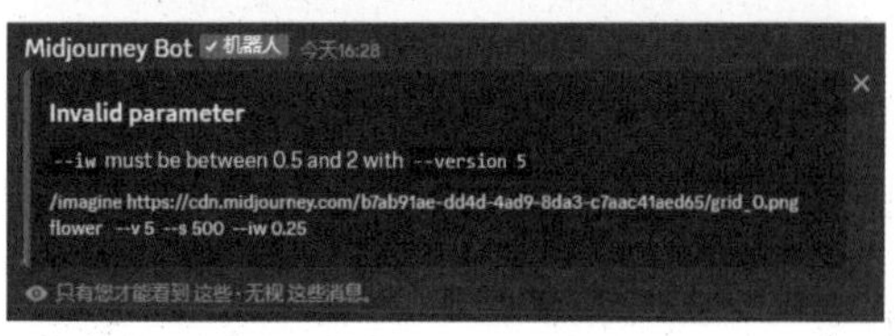

了解 Midjourney 的版本参数

MJ 虽然运行在网页端，但其实质仍然是一个软件，所以也具有软件版本号，只是与其他软件不同。不同版本的 MJ 并没有严格意义上的替代关系，因为不同版本的 MJ 擅长生成风格不同的图像。

V5.2 版本介绍

V5.2 版本于 2023 年 6 月 23 日发布，这一版本引进了全新的美学系统，可以生成更漂亮、清晰锐利的图像。此外，随着此版本还发布了几个新的功能，如 Zoom Out 按钮及 /shorten 命令。

V5.1 版本介绍

V5.1 版本于 2023 年 5 月 4 日发布，引入了 AI 自主理解功能，因此在图片生成方面更加贴近现实和用户的意图，对于生成广告、平面设计类图像来说有较大提升。与 V5 版本相比，此版本生成图像时，MJ 会自动添加符合提示语的细节。

V5 版本介绍

V5 版本擅长生成照片、写实类图像，且能生成分辨率更高的图像，如下面展示的人像。除非指定了图像类型，否则默认情况下使用此版本生成的图像大多数为照片。

V4 版本介绍

V4 版本在真实程度上稍逊于 V5 版本，尤其人的面部与手容易变形。但在图像的创意及艺术程度上高于 V5 版本，因此如果生成的是插画、科幻等图像，可以优先使用。

V3 版本介绍

V3 版本目前已经不建议使用，其特点是：与 V4、V5 版本相比，V3 版本生成的图像更加抽象，图像有较多杂色，且图像的整体性较低。下面是使用 V3 版本生成的图像。

但正是由于 V3 具有图像发散程度更高、更抽象的特性，因此可以用它来生成参考图，然后用本书讲解的以图生图的方式，使用 V4 或 V5 版本生成高质量图像，因此也不能说 V3 版本一无是处。

Niji 模型版本及参数介绍

了解 Niji 模型

Niji 是 MJ 专门用于生成插画类图像的模型，目前有两个版本，分别是 Niji version 4 与 Niji version 5，两个版本的区别及使用方法在本书第 6 章中有详细讲解。

无论使用哪一个版本的模型，均能够得到高质量插画图像，下面是使用 Niji version 5 生成的作品。

使用参数 --v 可以定义 V3、V4、V5 版本，如，--v 4、--v 5。

使用参数 --niji 4、--niji 5 可以定义 niji 的两个版本。

Niji 参数

如果使用的是 Niji 5 版本，则可以通过添加四个参数来控制生成图像的风格。

使用 --style cute 参数可以创作出更迷人可爱的角色、道具和场景。

使用 --style expressive 参数，可以让画面更精致更有表现力及插画感。

使用 --style original 参数，则可以让 MJ 使用原始 Niji 模型版本 5，这是 2023 年 5 月 26 日之前的默认版本。

使用 --style scenic 参数，可以让创作出来的画面更具奇幻的场景。

下面展示的是使用同样的描述语后，添加不同的参数获得的效果图。

Digital art, glitch art, web art, experimental art, cyber art, anime top model girl, collage --ar 2:3 --s 450 --style expressive --niji 5

Digital art, glitch art, web art, experimental art, cyber art, anime top model girl, collage --ar 2:3 --s 450 --style cute --niji 5

Digital art, glitch art, web art, experimental art, cyber art, anime top model girl, collage --ar 2:3 --s 450 --style original --niji 5

Digital art, glitch art, web art, experimental art, cyber art, anime top model girl, collage --ar 2:3 --s 450 --style scenic --niji 5

用 aspect 参数控制图像比例

可以用参数 --aspect 来控制生成图像的比例。在默认情况下，--aspect 值为 1:1，生成正方形图像。

如果使用的是 --v 5 版本，可以使用任意正数比例。

但如果使用的是其他版本，则需要注意比例的限制范围。对于 --v 4 版本，此数值仅可以使用 1:1、5:4、3:2、7:4、16:9 等比例值。

在实际使用过程中，--aspect 可以简写为 --ar。

--ar 1:2

--ar 9:16

--ar 3:4

--ar 1:1

--ar 4:5

--ar 2:3

用 quality 参数控制图像质量

在使用 MJ 时，可以使用参数 --quality 来控制生成图像的质量。较高的质量设置参数，需要更长的处理时间，但会产生更多细节。然而，较高的值也会消耗更多的 GPU 时间，因此会更消耗自己订阅的 GPU 时间量。

需要注意的是，较高的质量参数效果不一定更好，这取决于生成的图像的风格类型。例如，较低的 --quality 参数设置可能会更抽象外观，而较高的值可能会改善建筑、人像等需要更多细节的图像类型。

默认情况下，--quality 值为 1。如果使用的是 --v 5 及 --v 4 版本，则此数值的范围为 0.25 ~ 5。

在实际使用过程中，--quality 被简写为 --q，此参数设置不影响图像的分辨率。

在下面 3 张图中，左侧是值为 0.25 时的效果，中间是值为 2 时的效果，右侧是值为 5 时的效果。

对比 3 幅图像可以看出，图像的精细程度有明显差异。需要注意的是，当生成矢量化、块面化、细节较少的图像时，修改此数值得到的图像之间的区别并不大。

television, icon, white background,isometric --v 4 --q 5

television, icon, white background,isometric --v 4 --q 0.5

用 stylize 参数控制图像风格化

在使用 MJ 时，可以使用参数 --stylize 来控制生成图像的艺术化程度。较高的设置参数，需要更长的处理时间，但得到的效果更加艺术性，因此图像中有时会出现大量提示词没有涉及的元素，这也意味着最终得到的效果与提示词的匹配度更差。反之，越低的数值可使图像更加贴近提示词，但效果的艺术性也往往较低。

默认情况下，--stylize 值为 100。如果使用的是 --v5 及 --v4 版本，则此数值的范围为 100 ~ 1000。

在实际使用过程中，--stylize 被简写为 --s，此参数设置不影响图像的分辨率。

在下面的两排图像中，第一排参数为 1000，第二排参数为 100，这会导致图像艺术化差异明显。

Photograph taken portrait by Canon EOS R5,full body, A beautiful queen dress chinese ancient god clothes on her gold dragon throne ,Angry face, finger pointing forward,splendor chinese palace background， super wide angle,shot by 24mm les,in style of Yuumei Art, full portrait, 8k, photorealistic , elegant, hyper realistic, super detailed, portrait photography, global illumination --ar 2:3 --stylize 1000 --q 2 --v 5

Photograph taken portrait by Canon EOS R5,full body, A beautiful queen dress chinese ancient god clothes on her gold dragon throne ,Angry face, finger pointing forward,splendor chinese palace background， super wide angle,shot by 24mm les,in style of Yuumei Art, full portrait, 8k, photorealistic , elegant, hyper realistic, super detailed, portrait photography, global illumination --ar 2:3 --stylize 100 --q 2 --v 5

用 chaos 参数控制差异化

在使用 MJ 时，可以使用参数 --chaos 影响图像初始网格图的差异化程度。

较高的 --chaos 值会使 4 个网格图中的图像产生更大的区别，反之，使用较低的 --chaos 值，则会使 4 个网格图中的图像更加相似。

在默认情况下，此数值为 0。如果使用的是 --v 5 及 --v 4 版本，则此数值的范围为 0 ~ 100。在实际使用过程中，--chaos 被简写为 --c。

在下面的图像中，由于第二组使用了 --c 90 参数，因此 4 张图像之间有非常明显的差异。

natural lighting to highlight Persian Cat with whiskers with soft white curls ,full portrait shot of the cat , side view,action pose ,cat playing ball ,Use of a shallow depth of field to blur the background --q 2 --s 750 --ar 3:2 --v 5 --c 0

natural lighting to highlight Persian Cat with whiskers with soft white curls ,full portrait shot of the cat , side view,action pose ,cat playing ball ,Use of a shallow depth of field to blur the background --q 2 --s 750 --ar 3:2 --v 5 --c 90

用 repeat 参数重复执行多次生成操作

如果在 MJ 提示词后添加 --repeat 或 --r 参数，可以针对同样的提示词生成多组四格图像。例如，添加 --r 4，可以对提示词执行 4 次生成操作。

需要注意的是，针对不同等级的订阅用户，可以使用的数值范围不同。

针对标准用户，数值范围为 2 ~ 10。

针对 Pro 级用户，数值范围为 2 ~ 40。

建议在使用此参数时，配合前面讲解过的 --chaos 参数，这样就能快速生成大量可供选择的图像。

由于此命令会快速消耗订阅时间，因此执行时会弹出下图所示的提示。

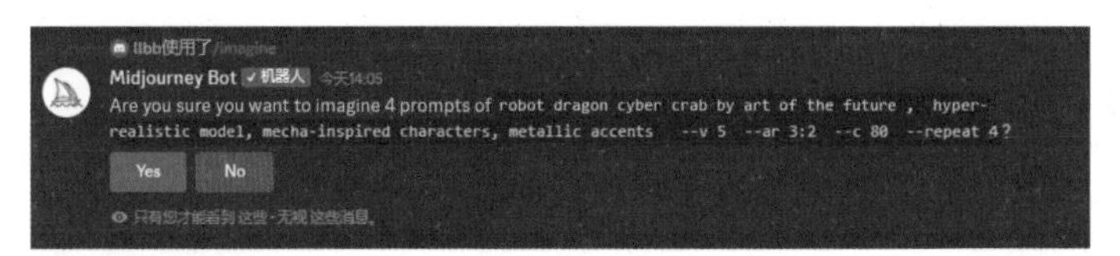

单击 Yes 按钮后，进入执行队列。

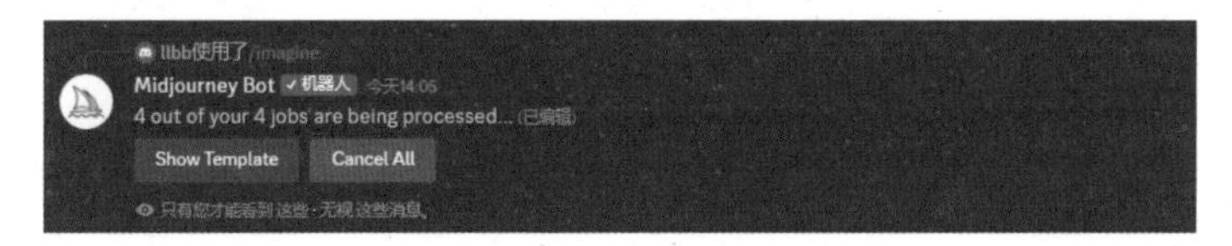

下面展示的是一级得到的 4 组效果，由于使用了 --c 80 参数，效果之间区别很大。

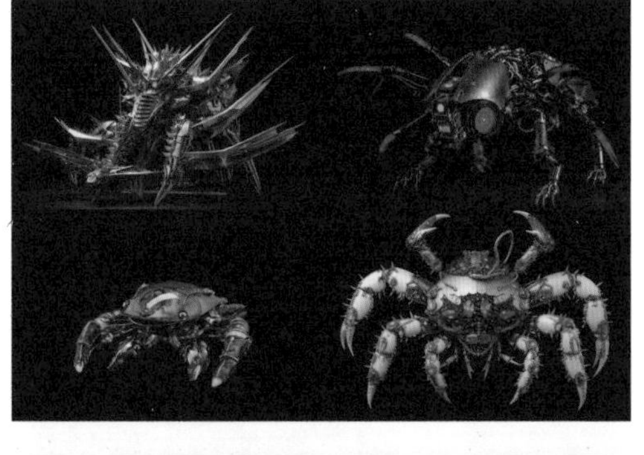

用 stop 参数控制图像完成度

如果在 MJ 提示词后添加 --stop 参数，则可以根据此参数的数值得到不同进度的生成图像。此参数的默认值为 100，意味着每次生成的图像完成度是 100%。

下面展示的是使用不同的 --stop 参数获得的图像。

这个参数并不常用，但如果对提示词生成的效果没有把握，为了节省订阅时间，可以使用 --stop 50 得到一张完成度为 50% 的图像，在观看此图像的基础上微调提示词。

当然，有时使用这一参数生成的未完成图像也恰好就是创作者需要的效果。

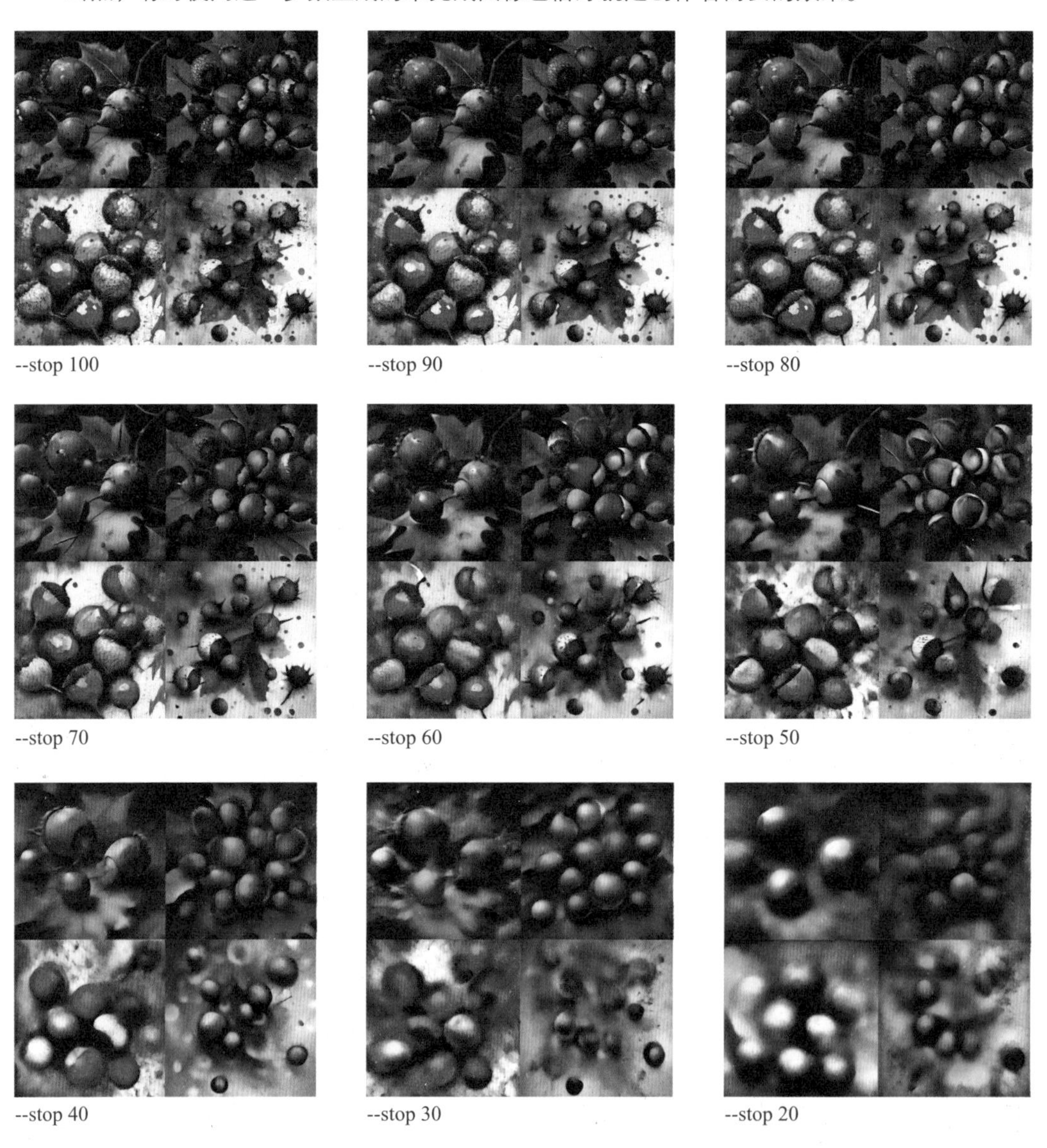

--stop 100　　--stop 90　　--stop 80

--stop 70　　--stop 60　　--stop 50

--stop 40　　--stop 30　　--stop 20

用 no 参数排除负面因素

如果不希望在生成的图像中包括某种颜色或元素，可以在 MJ 提示词后添加 --no 参数，然后添加针对性的负面词。

例如，针对 A girl smiled and reached out to receive a gift，a square-shaped wrapped box，clear background，vivid color，colorful --ar 3:2 --v 5 --c 10 --s 300 这一组提示词，生成的图像如左下图所示，如果不希望图像中有红色，则可以在这个提示词后面添加 --no red，这样生成的图像中就不会有红色，如右下图所示。

用 version 参数指定版本

如前所述，MJ 有多个版本，每个版本的算法各不相同，因此得到的图像风格也不同。在使用提示词时，可以在提示词后面添加 --version 或 --v 参数，从而为当前所要生成的图像指定不同的 MJ 版本。下面展示了同样一组参数分别使用 --v 5、--v 4、--v 3 得到的效果。

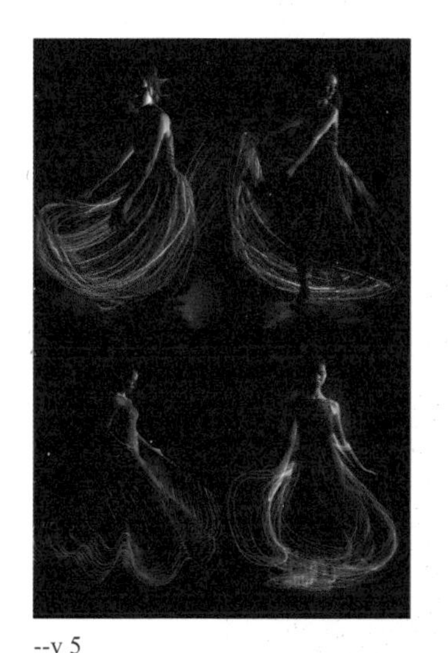

--v 5

--v 4

--v 3

用 seed 参数生成相同图像

MJ 在生成图像时，会使用一个 seed 数值来初始化原始图像，然后再根据这个原始图像利用算法逐步推演改进，直至得到创作者想要的图像。

seed 数值通常是一个随机数值，因此，如果不刻意使用此命令参数，即便用相同的提示词也不可能生成相同的图像，这也是在学习本书及其他提示词类教程时，即便创作者照搬照套提示词，也无法得到与示例图相同的图像的原因。

如果要得到相同的图像，可以为提示词指定同一 seed 值。

需要指出的是，创作者只能够获得自己曾经创作的作品的 seed 值，无法获得他人创作的 seed 值。

获得 seed 值方法

1. 在自己的创作界面中找到需要获得 seed 的作品，将光标放在提示词上，此时可以看到右侧有三个小点。

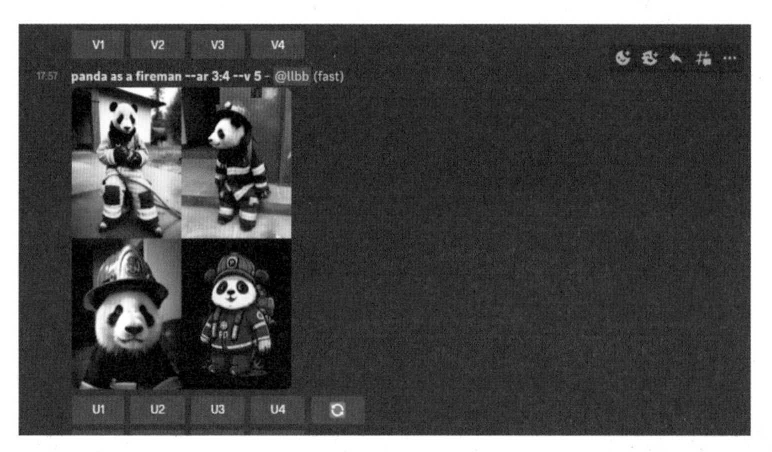

2. 单击三个小点后，选择“添加反应”命令，然后单击信封图标。

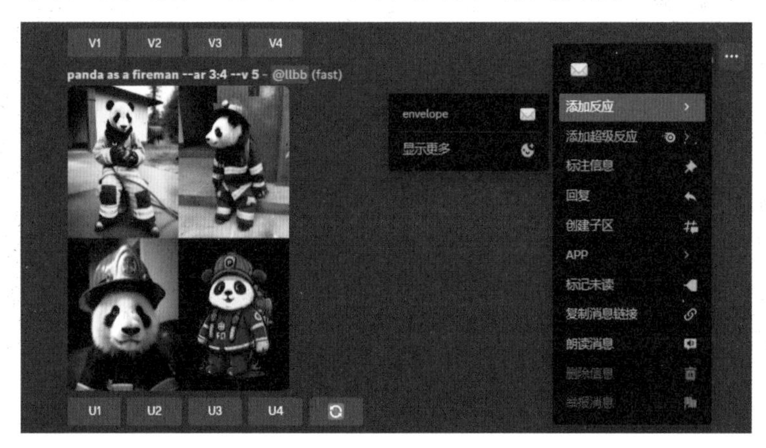

3. 单击右上角的收件箱图标。

4. 此时可以看到被查询作品的 seed 值。

5. 此时使用与被查询作品相同的提示词，再添加 --seed 命令，即可获得完全相同的图像。下方左图为原图像，右图为使用 --seed 值后生成的图像。

上面示例表明，如果希望他人生成相同的作品，只需要提交有 seed 值的相同提示词即可。同时也意味着要保护自己的图像版权，seed 值是必须保密的，因为除了 Pro 级订阅用户，其他 MJ 用户的作品的提示词均是公开的，所有 MJ 用户均可以看见并抄袭。

使用 --sameseed 参数获得类似图像

为了确保作品的多样性，使用 MJ 生成的四格图像通常拥有不同的构图风格。但如果希望四格图像彼此相近，生成有微妙变化的四格图像，则可以使用 --sameseed 命令。

需要注意的是，截至 2023 年 5 月 1 日，这个命令仅支持 V1、V2、V3 版本，无法在 V4、V5 版本中使用。

用 tile 参数生成无缝拼贴图案

无缝拼贴图案是由一个图像通过平移、旋转或翻转等操作得到的一组相似或完全相同的图像，这些图像可以拼贴在一起形成一个平铺的无缝图案。

无缝拼贴图案通常具备以下几个特点。

◎ 无缝衔接：相邻的图像衔接自然，没有明显的间隙或痕迹。

◎ 可重复：无缝拼贴图案可以重复无限次，形成一个无限大的平铺效果。

◎ 精确匹配：图像元素的颜色、亮度、对比度、纹理等应该相同或相似，以便拼贴出一个完美的图案。

要生成此类图案，可以在描述词后添加 --tile 参数。

例如，下面的提示词生成的是一种有浮雕效果花呢图案的墙纸无缝拼贴图案。

cream on light blue damask wallpaper, seamless tiled, half drop, embossed, --tile --v5

使用 MJ 生成此类图案后，得到的只是一个图案，如果要验证图案是否具有无缝拼贴效果，可以先将图案下载到本地，然后进入下方的网站。

https://www.pycheung.com/checker/

将保存在本地的图案直接拖至网站页面上，即可看到拼贴效果。

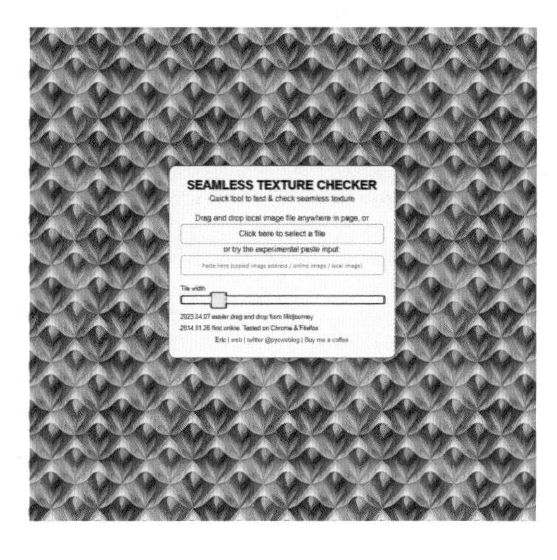

用 weird 参数生成古怪甚至诡异的图像

通过实验 --weird 或 --w 参数探索非传统美学。此参数为生成的图像引入了古怪和另类的品质，从而产生独特和意想不到的结果。

--weird 接受值：0~3000。

默认的 --weird 值为 0。

最佳 --weird 值取决于提示并需要进行实验。尝试从较小的值开始，例如 250 或 500，然后从那里向上 / 向下。如果您希望一代人具有传统的吸引力和怪异，请尝试将更高的 --stylize 价值观与 --weird. 尝试从两者相似的值开始。

--weird、--chaos 和 --stylize 之间是有区别，--chaos 参数控制初始网格图像彼此之间的差异程度，--stylize 控制了 MJ 的风格化美学的应用强度，而 --weird 控制了生成的图像与默认状态下 MJ 生成的符合常规美学定义的图像相比的异常程度。

acrylic art, art by Yoji shinkawa, kungfu character --ar 2:3 --s 750 --v 5.2

acrylic art, art by Yoji shinkawa, kungfu character --ar 2:3 --s 750 --v 5.2 --weird 200

acrylic art, art by Yoji shinkawa, kungfu character --ar 2:3 --s 750 --v 5.2 --weird 1500

acrylic art, art by Yoji shinkawa, kungfu character --ar 2:3 --s 750 --v 5.2 --weird 3000

掌握局部重绘功能

了解局部重绘

此功能是 MJ 于 2023 年 8 月 22 日发布的新功能，用于修改图像的局部。

长期以来，使用 MJ 生成图像都有一个比较大的问题，即图像效果的随机性较强，因此，有时得到的图像会有局部不能够令人满意的情况。在这种情况下，只能够将图像导入 Photoshop 中进行修改。

但使用局部重绘功能后，处理手段就会简单许多，只需要让 MJ 针对创作者指定的局部进行再次生成即可。

例如，左下图所示为有剑的原图，经过局部重绘后，可以在保持其他区域不变的情况下，抹除这把剑，如右下图所示。

局部重绘使用方法

使用局部重绘功能，需要先按 U 按钮放大图像，此时在图像下方将出现 Vary(Region) 按钮。

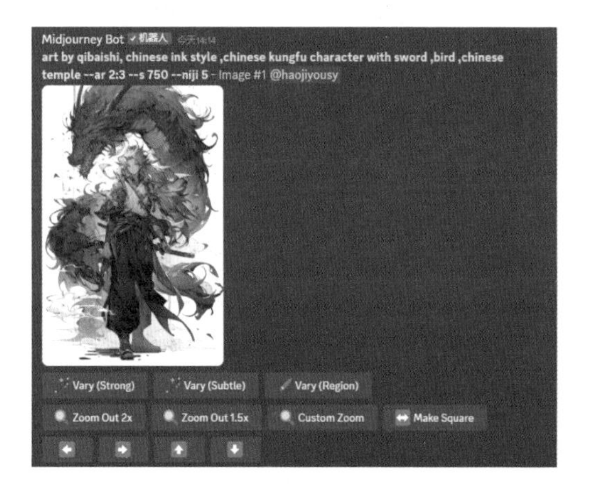

点击 Vary(Region) 按钮，进入局部重绘的界面，针对要重绘的局部拖出一个框，使其变为马赛克状态，则可完成局部重绘的区域定义。

然后在下方的描述语框中删除或修改字词，例如，在此笔者要去除剑，因此删除了对应的单词。

最后点击向右的箭头，即可得到新的四格图像。

在默认情况下，用于定义局部重绘区域的是方形选框工具，用于创建规则的选区，如果要重绘的区域是不规则形状，则要选择套索工具图标。

如果要撤销某一次操作，可以点击左上角的撤销图标。

局部重绘的使用技巧

使用此功能，可以完成修复局部元素、替换局部元素、删除局部元素、添加局部元素的任务。

在上面的示例中，笔者展示的就是删除局部元素的操作。

如果要添加局部元素，可以在画出要重绘的区域后，在提示语中添加要出现在图像中的描述词。

如果要修复局部元素，只需要画出对应的区域，保持描述语不变即可。例如，左下图所示为局部重绘前手的细节，右下图所示为重绘后手的细节。

第 2 章 掌握 Midjourney 提示语撰写逻辑及常用命令

认识提示语 Prompt 结构

在 MJ 中生成图像时，要在 /imagine 命令后面输入英文语句与参数，这些英文语句与参数可以统称为 Prompt，即提示语。

用好 MJ 的核心要点就是写出 AI 系统能理解的提示词，并确保提示词符合 AI 系统规范。

因此，要想用好 MJ，首先要了解提示词的结构，其次要掌握提示词的写作思路。

完整的 Prompt 分为三部分，即图片链接、文本提示词和参数。

图片链接

图片链接的作用是为 MJ 提供参考图，并影响最终结果，在本章后面介绍以图生成图的部分时会有详细讲解，下方的浅蓝色文字即为图片链接。

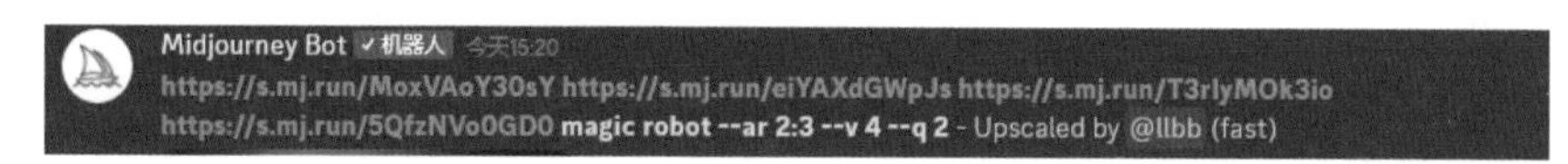

文本提示语

文本提示语是 MJ 的核心与学习重点，除非是采取以图片生成图片的方式进行创作，否则文本提示词是必不可少的部分。

根据要生成的效果，文本提示词可以简短为一个短句，如左下图所示，也可以复杂成为一篇小短文，如右下图所示。

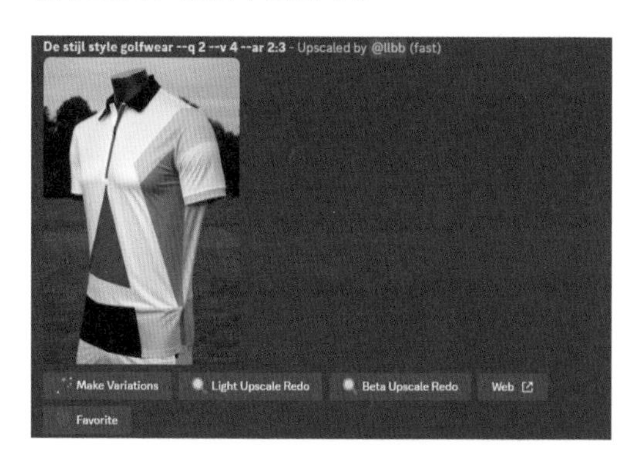

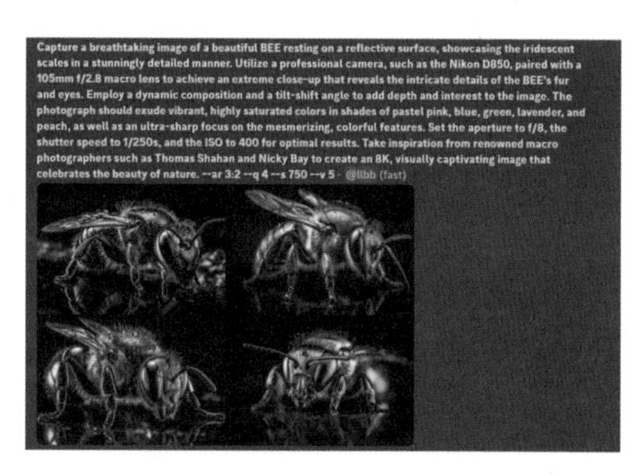

文本提示语是创作者需要关注的重点，也是本书讲解的重点，在后面的章节中，将分别讲解提示语的语法、撰写辅助工具、常用句式等内容。

参数

通常在每一个提示语的最后都要添加参数，以控制图像的生成方式，如宽高比、生成版本、质量等，不同的参数值对图像有不同的影响，这些参数在本章后面均有详细讲解。

例如，--ar 2:3 --q 5 --v 4 --c 50 --s 800 这一组参数定义了照片宽高比为 2:3，质量为 5，以 MJ v4 版本进行渲染生成，初始图像差异度为 50，风格化为 800。

利用翻译软件辅助撰写提示语

除非有深厚的英文功底，否则笔者建议创作者在撰写提示语时，打开 2 ～ 3 个在线翻译网站，先用中文描述自己希望得到的图像画面，再将其翻译成英文。

如果英文功底很弱，可以随便选择一个翻译后的文本填写在 /imagine 命令后面。

如果英文功底尚可，可以从中选择一个自己认为翻译得更加准确的文本填写在 /imagine 命令后面。

笔者经常使用的是百度在线翻译、有道在线翻译及 deepl 在线翻译。

下面是笔者给出的文本、翻译后的文本及使用此文本生成的图像。

两个维京武士军队相互进攻，在荒凉的平原上，雨水透过乌云向下倾盆而下。他们的旗帜在风中猎猎作响。一面旗帜上是黑乌鸦；另一面旗帜上是断裂的剑柄。在这片战场上，士兵们用力挥舞着手中的斧头和长剑相互厮杀，他们身上的盔甲在光芒中闪烁，他们的脸上写满了愤怒和威严。远处有火焰与浓烟。风暴席卷了整个战场，将相互攻击的士兵们的旗帜和长发吹得翻飞。雨水打湿了他们的盔甲和武器。一些士兵已经倒在了泥泞的地上。

two armies of viking warriors attacked each other, and rain poured down through dark clouds on a desolate plain. their flags hunted in the wind. on one banner, a black crow; on the other, a broken hilt. on this field, soldiers fought each other with axes and swords, their armor glinting in the light, their faces angry and majestic. there were flames and smoke in the distance.the storm swept across the field, blowing the flags and long hair of the soldiers who were attacking each other. the rain dampened their armour and weapons. some of the soldiers had fallen to the muddy ground.

从图像效果上来看，基本上达到了笔者心中构想的图像场景，在这个过程中，翻译软件起到了至关重要的作用。

使用 ChatGPT 辅助撰写提示语

ChatGPT 爆火可以说是最近几年来计算机领域现象级的事件，由于 ChatGPT 具有自动撰写文章及强大的翻译功能，因此，ChatGPT 与 MJ 结合进行创作，也自然成为许多同时掌握了 ChatGPT 与 MJ 使用方法的创作者的常用技巧。

要组合使用两者，需要先给 ChatGPT 设定任务背景，为此笔者使用了下面的设定命令。

bb　我正在使用Mj生成图像，Mj是一款人工智能图像生成软件，依靠提示语生成图像。我下面将提出初步概念，请将概念扩展成为一个复杂场景，并根据这个场景撰写一条完整的提示语，在这条提示语的描述中要包括对场景风格、视觉效果、光线、主题、气氛的描述，以确保Mj可以依靠这些提示语生成复杂的图像。你需要给我中英文对照的提示语，需要注意的是提示语要使用尽量简短的语句，尽量少用介词，如果明白了请回答明白，然后我将提出概念。

ChatGPT 给出了正确的反馈，笔者也得到了对应的提示语。

明白，我会尽力帮助您编写提示语。请提出您的初步概念，让我来为您扩展和撰写提示语。

bb　海底城市

中文提示语：在一座宏伟的海底城市中，高耸的建筑和错落有致的海藻缠绕在一起，构成了一个神秘而美丽的场景。海豚和鲨鱼在周围游动，阳光透过水面照射下来，形成柔和的光线，为整个城市增添了生机和神秘感。

英文提示语：In a grand underwater city, towering buildings and scattered seaweed intertwine to create a mysterious and beautiful scene. Dolphins and sharks swim around, and the sunlight shining through the water creates a soft light, adding vitality and mystery to the whole city.

接下来，笔者切换至 MJ 中，输入从 ChatGPT 中得到的提示语，并添加了参数，则得到了下方展示的图像，可以看出，效果还是非常好的。

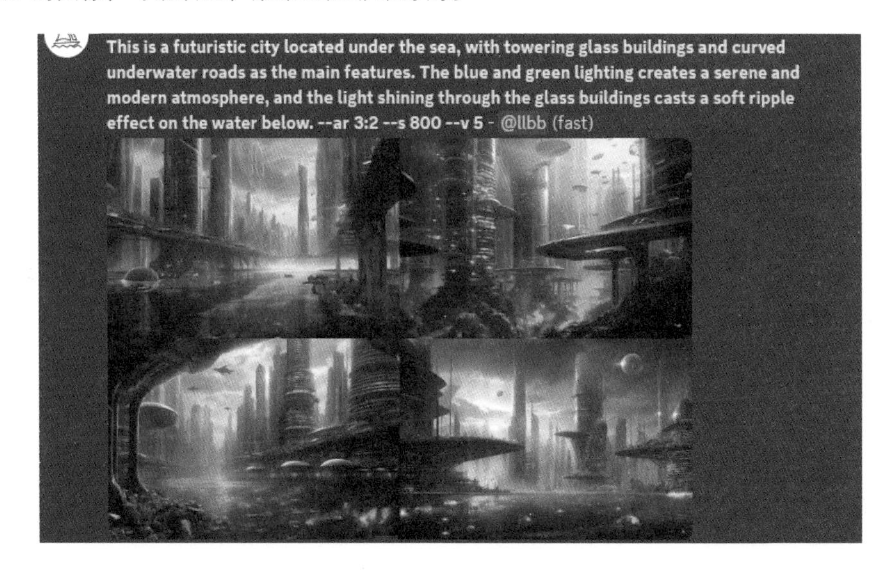

提示语中的容错机制

MJ 具有非常好的容错机制，包括拼写容错与语法容错，这意味着即便在撰写提示语的过程中使用了拼写错误的单词，或使用了错误的语法，也仍然能够得到正确的结果。

例如，下面的提示语 a beautiful chinese gril in red ancient costume is walking on a small path with a ancient chinaa village , holded a yellow umbrella. cloudy,backview,stones path.wide angle view,full body，分别在拼写、语法上出现了错误。

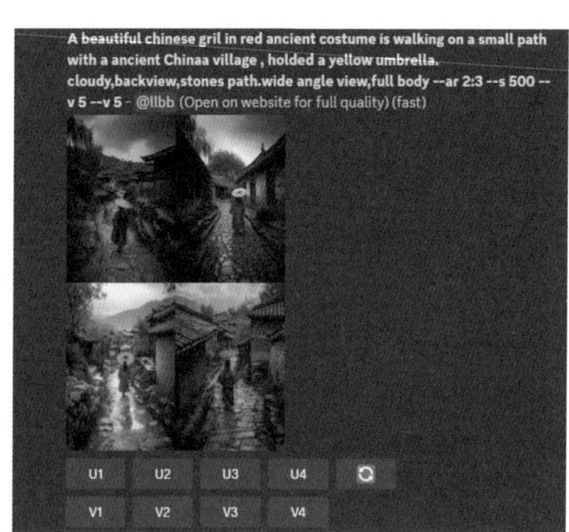

其中 gril 的正确拼写是 girl，with a ancient chinaa village 的正确语法与拼写是 in a ancient china village，holded 的正确时态应该是 holding，尽管出现了三种错误，但是从得到的图像来看，效果是正确的。

这意味着，对于英语基础一般的创作者来说，在保证总体正确性的基础上，不必过分在意语法与不会产生歧义情况下的单词拼写。

提示语中的违禁词规范

与国内的媒体平台不允许使用许多违禁词一样，使用 MJ 生成图像时，也要注意避免使用与人体隐私、政治、宗教、血腥等负面内容相关的词汇。

目前，MJ 并没有发布违禁词列表，但如果提示语中不小心加入了这样的词汇，MJ 会出现相关提示。例如，当笔者在提示语中加入了 blood、bloody 时，则会自动触发 MJ 的审核机制，弹出如左下图所示的错误提示信息。

如果多次触发这样的提示信息，则账号会被临时禁用，直至被官方人员手工解除禁用限制，如右图所示。

Midjourney Bot ✓机器人 今天11:55

Banned prompt

The word bloody is banned. Circumventing this filter to violate our rules may result in your access being revoked.

/imagine Viking warriors gathered on the desolate plains as the black sky poured down through the dark clouds. Their banners hunted in the wind in anticipation of the coming battle. On one banner were beak-shaped heads that hissed terribly; on the other were broken sword hilts and blood-stained blades, foreshadowing the bloody cruelty to come. The battlefield was filled with the scent of death and pain, and the sound of swords and human cries intertwined to form a stern music. On this battlefield, the soldiers wielded their axes and longswords with great force, their chain armor and iron helmets shining in the light, presenting a magnificent and terrible sight. Their faces were written with anger and majesty as they fought forward for the honor and glory of their clan. The storm swept across the battlefield, blowing the banners and long hair of the soldiers over and over. The rain drenched their armor and weapons, making the battle even more gruesome and brutal. Some soldiers had already fallen to the muddy ground, their blood seeping into the ground and forming a dark red furrow.epic , --ar 9:16 --s 850 --q 3 --v 5

Nsfw alert

You have triggered an abuse alert. Your account is now under manual review. You will be unblocked 40秒前

/imagine Viking warriors gathered on the desolate plains as the black sky poured down through the dark clouds. Their banners hunted in the wind in anticipation of the coming battle. On one banner were beak-shaped heads that hissed terribly; on the other were broken sword hilts and stained blades, . The battlefield was filled with the scent of death and pain, and the sound of swords and human cries intertwined to form a stern music. On this battlefield, the soldiers wielded their axes and longswords with great force, their chain armor and iron helmets shining in the light, presenting a magnificent and terrible sight. Their faces were written with anger and majesty as they fought forward for the honor and glory of their clan. The storm swept across the battlefield, blowing the banners and long hair of the soldiers over and over. The rain drenched their armor and weapons, making the battle even more gruesome and brutal. Some soldiers had already fallen to the muddy ground, .epic , --ar 9:16 --s 850 --q 3 --v 5

掌握提示语大小写及标点用法

由于 MJ 是一个运行于英文平台的 AI 系统，因此提示语遵从英语语法结构。下面是笔者在创作中总结出来的一些语法要点。

提示语大小写及标点符号规范

提示语是用英文撰写的，在撰写时并没有太多规范，既可以使用大写也可以使用小写，还可以大小写混用，可以使用英文句号或逗号，也可以使用中文句号或逗号。但在撰写参数时，必须使用英文符号，而且要注意使用空格。

正确运用标点符号，可以帮助 MJ 更正确地理解提示语所要表达的意思。

下面三组图像使用的是完全相同的提示语，区别在于，左侧图像没有添加任何标点符号，中间的图像添加了正确的标点符号，而右侧的图像添加了错误的标点符号。

对比可见，第三组使用了错误的标点符号，导致图像出现明显错误（花出现在雕塑的身上），效果甚至不如完全没有标点的第一组图像。

第一组图像虽然没有添加标点符号，但 MJ 也“猜”到了笔者想要的效果，这表明对于容易理解的语句及常见的单词，MJ 依据大数据是可以“猜”对的。

feminine lakshmi with elegant flowers surrouiding her below side view with colored detailing gold edges indian details hyperrealistic extremely realistic cinematric lights global illumination volumetric lights shot by canon dslr wide angle --ar 2:3 --v 5 --q 2 --s 800

feminine lakshmi with elegant flowers surrouiding her, below side view with colored detailing gold edges, indian details, hyperrealistic extremely, realistic cinematric lights, global illumination, volumetric lights, shot by canon dslr, wide angle --ar 2:3 --v 5 --q 2 --s 800

feminine lakshmi, with elegant flowers, surrouiding her, below side view, with colored detailing, gold edges, indian details, hyperrealistic extremely, realistic, cinematric lights, global illumination, volumetric lights, shot by canon dslr, wide, angle --ar 2:3 --v 5 --q 2 --s 800

另外，在提示语中除了双冒号（::）外，其他的标点符号只起分隔作用。右图所示为笔者分别将逗号换成感叹号与括号后的图像，从效果可以看出，这两组图像与上一页中使用逗号得到的图像从效果上看没有多少区别。

feminine lakshmi with elegant flowers surrouiding her , below side view with colored detailing gold edges! indian details! hyperrealistic extremely !realistic cinematric lights! global illumination! volumetric lights! shot by canon dslr!wide angle! --ar 2:3 --v 5 --s 800

(feminine lakshmi with elegant flowers surrouiding her)(below side view with colored detailing gold edges)(indian details)(hyperrealistic extremely)(realistic cinematric lights)(global illumination)(volumetric lights)(shot by canon dslr)(wide angle) --ar 2:3 --v 5 --s 800

利用双冒号控制文本权重

在使用提示词生成图像时，除非提示词非常简单，否则在一个完整的提示语中都会出现多个控制最终图像的文本短句，利用英文双冒号，可以有效控制不同文本短句对图像的影响程度，即改变文本的权重。

例如，针对提示语 movie poster design, a pretty chinese girl is charging forward with a sword in hand, snowy weather, petals falling, full body, dynamic pose, the sword shining（电影海报设计，一个美丽的中国女孩手持剑向前冲刺，下雪天，空中飘落着花瓣，女孩的全身呈现出动感的姿态，手中的剑闪烁着光芒），可以拆解为以下六小段控制最终图像的文本短句。

◎ movie poster design（电影海报设计）

◎ a pretty chinese girl is charging forward with a sword in hand（一个美丽的中国女孩手持剑向前冲刺）

◎ snowy weather（下雪天）

◎ petals falling（空中飘落着的花瓣）

◎ full body dynamic pose（女孩的全身呈现出动感的姿态）

◎ the sword hining （手中的剑闪烁着光芒）

如果不干涉各文本的权重，则 MJ 默认所有权重为 1，生成的图像如下图所示。

如果在各文本短句中添加控制文本权重的英文双冒号，则可使 MJ 更突出某一细节。

下面笔者先展示一个极端的案例，在 snowy weather 与 petals falling 后面分别添加了英文双冒号及较大权重值，使提示语变为 movie poster design, a pretty chinese girl is charging forward with a sword in hand, snowy weather :: 20 petals falling::30, full body, dynamic pose, the sword shining. --ar 2:3 --s 800 --v 5，此时得到的图像如下图所示，可以看出，由于其他段落的权重默认为 1，而 petals falling 为 30，使图像仅突出了空中飘落花瓣的效果。

下面笔者将提示语修改为 movie poster design::30, a pretty chinese girl is charging forward with a sword in hand, snowy weather::4 petals falling::2 , full body,dynamic pose,the sword shining . --ar 2:3 --s 800 --v 5，在这个提示语中，海报设计权重被提高，因此得到了如右图所示的效果。

必须掌握的常用提示关键词

虽然，可以使用 MJ 生成千变万化的图像，撰写出来的提示语也各不相同，但其中仍然有一些提示语关键词的使用频率较高。类似于英语中的高频词，只要掌握了这些高频词，就能应对大部分提示语的撰写任务。

下面是笔者经常使用的一些提示语高频词。

控制材质的关键词

当要控制生成的图像中对象的材质时，可以使用关键词 made of ……，在 of 后面可以添加任何对象，例如 fly dragon made of electronic components and pcb circuits，使用的材质是电子元件和 pcb 电路板。

例如，按照这个句式，可以生成下方使用四种材质制作的鞋子图像。从左到右材质依次为 gold glitter payette（金色闪闪发光的珠片）、mother of pearl and diamonds（珍珠母贝和钻石）、shining diamonds（发光的钻石）、 lace and ribbons（蕾丝与丝带）。

控制风格的关键词

风格控制是撰写提示语的重要步骤，可以使用 in style of……句式，在 of 的后面可以添加各种风格，如中式、欧式、维多利亚风格，也可以添加各个艺术家的名称，如凡·高、毕加索等，还可以添加知名的 ip 形象，如钢铁侠、星球大战等。

使用……style 句式，并在 style 前面添加关键词可以起到同样的作用。

此外，也可以用 by、design by 句式，在 by 的后面添加风格控制关键词。

例如，按照以上句式，可以生成下方三种不同风格的相机。从左到右使用的风格关键词依次为 rococo style（洛可可风格）、star war style（星球大战风格）、optimus prime style（变形金刚擎天柱风格）。

控制图像主题的关键词

使用 the theme of 及 themed 可以让 MJ 明白图像的主题，并在生成图像时添加相对应的元素。

例如，a harry potter-themed birthday party，指的是以哈利·波特为主题的生日派对。

a winter wonderland-themed wedding，指的是以冬季仙境为主题的婚礼。

a superhero-themed amusement park，指的是以超级英雄为主题的游乐园。

以上三句提示，也可以使用 the theme of 表达。

a birthday party with the theme of harry potter

a wedding with the theme of winter wonderland

an amusement park with the theme of superheroes

下面是 a superhero-themed amusement park 生成的游乐园图像，其中明显有许多超级英雄类电影中的元素及符号。

控制背景的关键词

使用 MJ 生成图像，有时需要控制生成图像的背景，如白色、黑色、灰色等，此时可以使用…… background 句式。在 background 前面添加颜色或其他词汇，如 future city background（未来城市背景）、grassland background（草原背景）、 street background（街头背景）、lush forest background（树林背景）。

控制元素数量的关键词

有时可能需要在提示语中添加数量，如5个苹果、9个人等，虽然根据笔者测试，到目前为止，MJ 尚无法精确控制图像中元素的数量。但这并不意味着，写入数量肯定无法得到正确的图像。

例如，在生成左下图所示的图像时，笔者使用的提示语为 there are three girls in the classroom（教室里有 3 个女孩），生成中间图像时使用的提示语为 there are five white doves flying in the square（广场上飞舞着 5 只白鸽），生成右侧图像时使用的提示语为 there are nine girls in the classroom,photo（教室里有 9 个女孩）。

很明显，这三张图像一正两误，这意味着当在提示语中添加控制元素数量的句式时，得到的结果具有随机性。

there are three girls in the classroom

there are five white doves flying in the square.

there are nine girls in the classroom

但MJ能够较好地处理不太精确的数量，例如可以使用 few（很少的，少数的）、several（几个）、many（许多，很多）、numerous（大量的）、a couple of（两个，几个）、dozens of（几十个）、scores of（许多）、hundreds of（数百个）、thousands of（数千个）、large pile of （一大堆的）等关键词，在图像中展示相对正确的非精确元素量级。例如左下方的图像使用了 large pile of（一大堆的）、右下方的图像使用了 few（很少的，少数的），得到的图像中元素的量级控制是正确的。

large pile of gold jewelry --ar 3:2 --v 5 --s 750 --q 2

few gold jewelry --ar 3:2 --v 5 --s 750 --q 2

撰写提示语的 4 种方法

关键词随机联想法

如果创作者对于希望生成的图像并没有精确的要求，则可以使用这种随机联想法，这种方法只要求创作者提供关于图像的关键词，将这些关键词罗列出来即可，创作者不必撰写出符合语法的长句。

这样做的好处是，如果能够提供足够多的关键词，生成的图像与心中所构思的图像相差不会太大。

例如，笔者以雾、高山、峡谷、桃花、小溪、夕阳、行者、晚霞、飞鸟（mist, high mountains, canyons, peach blossoms, streams, sunset, traveler, evening glow, flying birds.）为关键词生成的图像如下图所示，与心中所构思的场景区别不大。

由于希望生成的是照片类图像，因此在提示语中添加了 shot by max rive（ 由 max rive 拍摄），max rive 是一位知名的风景摄影师，他的作品以壮观的自然风景和精美的摄影技术著称。

如果希望生成不同类型的图像，可以只撰写关键词，并注意删除关于图像类型的描述。

例如，笔者以办公室、忙碌、女职员、简洁、高科技、计算机、白色（office， busy， female staff， simple， high-tech， computer， white）为关键词生成的图像如下图所示，其中有两张照片类型的图像，也有简笔画与矢量插图类型的图像。

通用词刷新迭代法

这种方法是指当创作者对于图像的细节没有具体要求，只是模糊地知道方向，则可以利用通用词来进行创作，然后不断点击 v 按钮，直至刷出满意的效果。

例如，要生成一个王冠的图像，可以直接用 a kings crown 这个关键词，除非对王冠的造型、颜色有特别的要求，否则在点击 v 按钮两三次后，总能够获得还不错的图像。

实际上，在大多数情况下，创作者对图像都不会有特别精确的要求，因此，在创作一张图像时完全可以先通过这种方法，先看看是否能够得到令人满意的作品，如果其中有还算不错的图像，则可以利用本书后面将讲解的，使用 Photoshop 的 AI 功能对图像细节进行编辑加工的方法来处理图像，直至得到成品。如果掌握了 Stable Diffusion，也可以利用其对细节进行重绘。

提示语网站参考法

对于初学者来说，学习的最好方式之一莫过于模仿。在撰写提示语前，前往以下网站搜索与自己希望创作的图像主题有关的关键词，即可获得大量有提示语的参考图像，有时在某一张参考图像的提示语的基础上，只需要稍微修改几个关键词就能够获得令人满意的效果。

https://www.midjourney.com/app/feed/?sort=new

https://lexica.art/

https://prompthero.com/midjourney-prompts

https://qfbz1.cn/

图像细节描述法及常用关键词

这是最常用的一种方法，即在提示语中详细描述要生成的图像的主要细节。

描述时可以参考下面这个通用模板。

主题 、主角、 背景、环境、气氛、构图、镜头、风格化、参考、图像类型

这个模板的组成要素解释如下。

◎ 主题内容：要描述出想要绘制的主题，如珠宝设计、建筑设计、贴纸设计等。

◎ 主角：既可以是人也可以是物，对其大小、造型、动作等进行详细描述。

◎ 环境：描述主角所处的环境，如室内、丛林中、山谷中等。

◎ 气氛：包括光线，如逆光、弱光，以及天气，如云、雾、雨、雪等。

◎ 构图：描述图像的景别，如全景、特写等。

◎ 风格化：描述图像的风格，如中式、欧式等。

◎ 参考方向：描述生成图像时 MJ 的参考类型，可以是艺术家名称，也可以是某些艺术网站。

◎ 图像类型：包括图像是插画还是照片，是像素画还是 3D 渲染效果等信息。

在具体撰写时，可以根据需要选择一个或几个要素来进行描述。

同时需要注意的是，避免使用没有实际意义的词汇，如画面有紧张的气氛、天空很压抑等。

最后，建议使用简短的小句子，而避免使用大量介词构成的长句。

下面笔者通过分析一个提示语来展示具体应用。

the girls stand on a street corner, one dressed in trendy, streetwear-inspired clothes while the other dons flowy, bohemian attire. the scene features a mix of natural and artificial light, with buildings and cityscape visible in the background. wide angel full portrait --ar 2:3 --s 600 --v 5（女孩们站在街角，一个身穿时尚的街头风装扮，另一个穿着飘逸的波希米亚服饰。背景中可见建筑和城市景观，自然光和人工光混合，广角，全身照）

在上面的提示语中，主角是 the girls，动作描述是 stand ，环境是 a street corner 及 with buildings and cityscape visible in the background，主角造型是 one dressed in trendy, streetwear-inspired clothes while the other dons flowy，气氛是 mix of natural and artificial light，构图是 wide angel full portrait，图像类型由参数 --v 5 确定为照片类型。

在使用这种方法描述具体图像时，通常要使用到下面列举的景别、视角、光线、情绪、天气、环境、材质等方面的关键词。

景别关键词

特写 close-up、中特写 mediumclose-up、中景 mediumshot、中远景 mediumlongshot、远景 longshot、背景虚化 bokeh、全身照 fulllengthshot、大特写 detailshot、腰部以上 waistshot、膝盖以上 kneeshot、脸部特写 faceshot。

视角关键词

广角视角 wide angle view、全景视角 panoramic view、低角度视角 low angle shot、俯拍视角 overhead、常规视角 eye-level、鸟瞰视角 aerial view、鱼眼视角 fisheye lens、微距视角 macrolens、顶视图 top view、称轴视角 tilt-shift、卫星视角 satellite view、底视角 bottom view、前视 front view、侧视 side view、后视 back view。

光线关键词

体积光 volumetric lighting、电影灯光 cinematic lighting、正面照明 front lighting、背景照明 back lighting、边缘照明 rim lighting、全局照明 global illuminations、工作室灯光 studio lighting、自然光 natural light。

情绪关键词

愤怒 angry、高兴 happy、悲伤 sad、焦虑 anxious、惊讶 surprised、恐惧 afraid、羞愧 embarrassed、厌恶 disgusted、惊恐 terrified、沮丧 depressed。

天气关键词

晴天 sunny、阴天 cloudy、雨天 rainy、下雨 rainy、暴雨 torrentialrain、雪天 snowy、小雪 lightsnow、大雪 heavysnow、雾天 foggy、多风 windy。

环境关键词

森林 forest、沙漠 desert、海滩 beach、山脉 mountain range、草原 grassland、城市 city、农村 countryside、湖泊 lake、河流 river、海洋 ocean、冰川 glacier、峡谷 canyon、花园 garden、森林公园 national park、火山 volcano。

材质关键词

木头 wood、金属 metal、塑料 plastic、石头 stone、玻璃 glass、纸张 paper、陶瓷 ceramic、丝绸 silk、棉布 cotton、毛料 wool、皮革 leather、橡胶 rubber、珍珠 pearl、大理石 marble、珐琅 enamel、绸缎 satin、细麻布 linen、纤维素 cellulose、金刚石 diamond、羽毛 feather。

利用提示语中的变量批量生成图像

单变量的使用方法

在最新的 MJ 版本中，可以使用类似于编程中排列组合各类变量的方法来批量生成图像。

基本方法是将参数变量放在 {} 中，并以逗号进行分隔。

例如，当创作者撰写并执行 a naturalist illustration of a {pineapple, blueberry, rambutan, banana} bird 这样一个提示语时，实际上 MJ 将会把这条提示语分解为以下 4 条，从而生成 4 组四格初始图像。

a naturalist illustration of a pineapple bird

a naturalist illustration of a blueberry bird

a naturalist illustration of a rambutan bird

a naturalist illustration of a banana bird

可以看出来，这样的命令格式大大加快了创作者生成图像的效率，当然也会大大加快创作者消耗订阅时间的速度。

除了可以在提示语的文本段落中使用变量，还可以将参数当作变量使用。

例如，当创作者撰写并执行 a bird --ar {3:2, 1:1, 2:3, 1:2} 这样一个提示语时，实际上 MJ 将会把这条提示语分解为以下 4 条，从而生成 4 组内容相同但画幅比例不同的四格初始图像。

a bird --ar 3:2

a bird --ar 1:1

a bird --ar 2:3

a bird --ar 1:2

同理，也可以将 --s、--v、--c、--q 等参数当作变量加到提示语中。

多变量的使用方法

在一组提示语中，可以使用多个变量。

例如，当创作者撰写并执行 a {bird,dog} --ar {3:2, 16:9} --s {200，900} 这样一个提示语时，实际上，MJ 将会把这条提示语分解为以下 8 条，从而生成 8 组内容不同、画幅比例不同、风格不同的四格初始图像。

a bird --ar 3:2 --s 200

a bird --ar 16:9 --s 200

a bird --ar 3:2 --s 900

a bird --ar 16:9 --s 900

a dog --ar 3:2 --s 200

a dog --ar 16:9 --s 200

a dog --ar 3:2 --s 900

a dog --ar 16:9 --s 900

嵌套变量的使用方法

前面列举的都是单变量使用实例，根据需要还可以使用更复杂的嵌套变量。

例如，当创作者撰写并执行 a {sculpture, painting} of a {apple {on a pier, on a beach}, dog {on a sofa, in a truck}}. 这样一个提示语时，实际上 MJ 将会把这条提示语分解为以下 8 条，从而生成 8 组不同的四格初始图像。

a sculpture of a appleon a pier.

a sculpture of a apple on a beach.

a sculpture of a dog on a sofa.

a sculpture of a dog in a truck.

a painting of a apple on a pier.

a painting of a apple on a beach.

a painting of a dog on a sofa.

a painting of a dog in a truck.

以下为对变量组合的拆解。

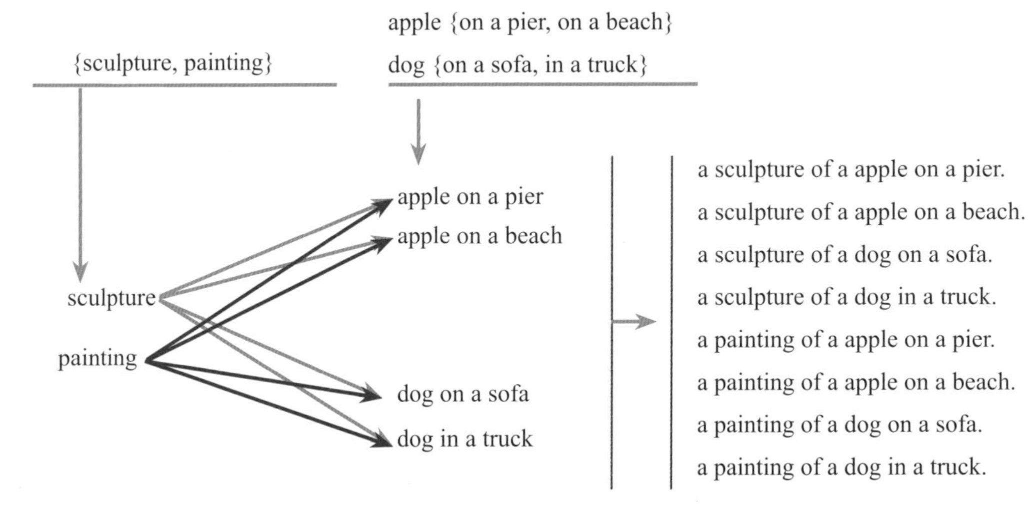

使用变量撰写提示语的方法特别适合于完成形式固定、图像各异的任务。例如，要为涂色书生成一批图像，则可以使用提示语 coloring page for adults, clean line art, {dog，cat，elephant，tiger，lion，bear，deer，giraffe，monkey，penguin，dolphin，whale，kangaroo，crocodile，snake} --v 4 --s 750。

这样就能一次性针对 15 种动物生成 60 幅涂色图像，从而能够极大地提高效率。

以图生图的方式创作新图像

基本使用方法

MJ 具有很强的模仿能力，可以使用图像生成技术生成类似于原始图像的新图像。这种技术使用深度学习神经网络模型来生成具有相似特征的图像。

在图像生成中，神经网络模型通常被称为生成对抗网络（generative adversarial network，gan），由生成器（generator）和判别器（discriminator）两个神经网络组成。生成器负责生成新图像，而判别器负责识别生成器生成的图像是否与真实图像相似。这两个神经网络不断互相对抗和学习，使得生成的图像逐渐接近创作者上传的参考图像。

使用步骤如下所述。

1. 单击命令行中的 + 号，在菜单中选择“上传文件”命令，然后选择参考图像。

2. 图像上传完成后，会显示在工作窗口。

3. 选中这张图像，然后单击鼠标右键 ，在弹出的快捷菜单中选择“复制图片地址”命令，然后单击其他空白区域，退出观看图像状态。

4. 输入或找到 /imagine 命令，在参数区先按【Ctrl+V】组合键执行粘贴操作，将上一步复制的图片地址粘贴到提示词最前方，然后敲空格键，输入对生成图片效果、风格等方面的描述，并添加参数，敲回车键确认，即可得到所需的效果。

左下图所示为笔者上传的参考图像，中图所示为生成的四格初始图像，右图所示为放大其中一张图像后的效果，可以看出，整体效果与原参考图像相似，质量不错。

图生图创作技巧 1——自制图

使用图生图时，一个有用的技巧是自制参考图，这需要有一定的 Photoshop 软件应用技巧，但却可以得到更符合需求的参考图。创作者可以根据自己的想象，将若干个元素拼贴在一张图中，操作时无须考虑元素之间的颜色、明暗匹配关系，只需考虑整体构图及元素比例即可。

例如，左图所示为笔者使用若干元素拼贴而成的一张参考图，可以明显看出，各个元素之间的颜色与明暗有很大差异。

中图所示为根据此参考图得到的四张初始图像，右图所示为放大后的效果。

下面展示一些笔者使用这种自制图方法制作的示例。

图生图创作技巧 2——多图融合

在前面的操作示例中，笔者使用的都是一张图，但实际上，创作者可以根据需要使用多张图像执行图像融合操作。

但操作方法与使用一张图并没有不同，区别在于需要上传 2 张以上的图像。

如果希望控制图像融合的效果，可以在提示词中图片地址的后方输入希望生成的图像效果及风格，如果只希望简单融合图像，可以只输入参数值。

例如，在创作下面的两组图像时，笔者都只输入了参数值，因此最终融合得到的图像是由 MJ 平衡地提取了参考图像中最典型的特征后生成的。

例如，第一组图像中左侧参考图的武器、长发，中间图像的齿轮、服装，均很融洽地出现在最终的融合图像中。

由于两张参考图像彼此相差较大，因此，第二组图像虽然最终的效果图也能明显看出两张图片的特征，但整体效果比较出乎意料。这也提示创作者，在融合时最好不要使用完全不相同的图，或者注意在提示词中添加关键词以对效果进行控制。

图生图创作技巧 3——控制参考图片权重

当用前面讲述的以图生图的方法进行创作时，可以用图像权重参数 --iw 来调整参考图像对最终效果的影响效果。

较高的 --iw 值意味着参考图像对最终结果的影响更大。

不同的 MJ 版本模型具有不同的图像权重范围。

对于 v5 版本，此数值默认为 1，数值范围为 0.5 ~ 2。对于 v3 版本，此数值默认为 0.25，数值范围为 -10000 ~ 10000。

右图所示为笔者使用的参考图，提示语为 flower --v 5 --s 500，下面 4 张图为 --iw 参数为 0.5（左上）、1（右上）、1.5（左下）、2（右下）时的效果图。

通过图像可以看出来，当 --iw 数值较小时，提示语 flower 对最终图像的生成效果影响更大；但当 --iw 数值为 2 时，生成的最终图像与原始图像非常接近，提示语 flower 对最终图像的生成效果影响不大。

用 blend 命令混合图像

/blend 是一个非常有意思的命令，当创作者上传 2 ～ 5 张图像后，使用此命令可以将这些图像混合成一张新的图像，这个结果有时可以预料，有时则完全出乎意料。

基本使用方法

1. 在命令行中找到或输入 /blend 后，则 MJ 显示如下图所示的界面，提示创作者要上传两张图像。

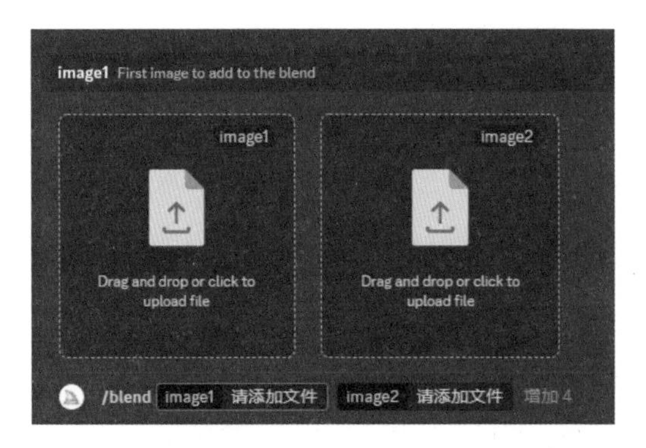

2. 可以直接通过拖动的方法将两张图像拖入上传框中。下图就是笔者上传图像后的界面。

3. 在默认情况下，混合生成的图像是正方形的，但创作者也可以自定义图像比例，方法是在命令行中单击一下，此时 MJ 会显示更多参数，其中 dimensions 用于控制比例。

4. 选择 dimensions 后，可以选择 portrait、square、landscape 3 个选项，其中 portrait 生成 2:3 的竖画幅图像，square 生成正方形图像，landscape 生成 3:2 的横画幅图像。

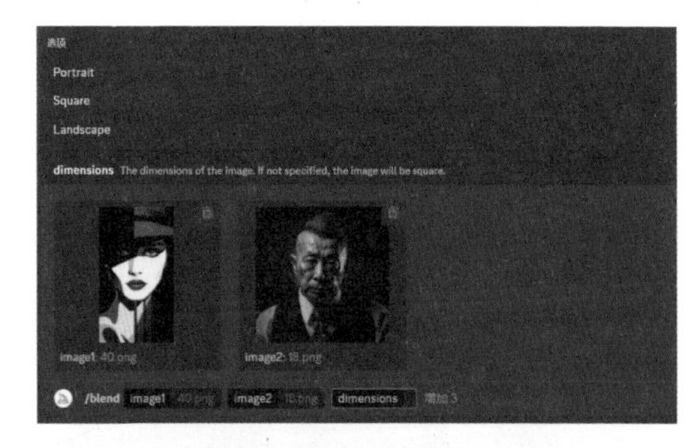

5. 按回车键后，则 MJ 开始混合图像，得到如右侧图所示的效果图。

混合示例

可以尝试使用 /blend 命令混合各类图像，以得到改变风格、绘画类型、颜色等元素的图像，下面是一些示例，左侧两图为原图，右侧两图为混合后的效果。

使用注意事项

使用 /blend 命令混合图像的优点是操作简单，缺点是无法输入文本提示词。因此，如果希望在混合图像的同时，还能够输入自定义的提示词，应该使用前面所讲述的 /imagine 命令，通过上传图像后获得的图像链接地址进行混合的方法。

用 describe 命令自动分析图片提示词

MJ 的一大使用难点就是撰写准确的提示词，这要求使用者有较高的艺术修养与语言功底，针对这一难点 MJ 推出了 describe 命令。

使用这一命令，可以让 MJ 自动分析使用者上传的图片，并生成对应的提示词。虽然每次分析的结果可能并不完全准确，但大致方向并没有问题，使用者只需在 MJ 给出的提示词基础上稍加修改，就能够得到个性化的提示词，进而生成令人满意的图像。

下面是基本使用方法。

1. 找好参考图后，在 MJ 命令行处找到 /describe 命令，此时 MJ 将出现一个文件上传的窗口。

2. 将参考图直接拖到此窗口以上传此参考图，然后按回车键。

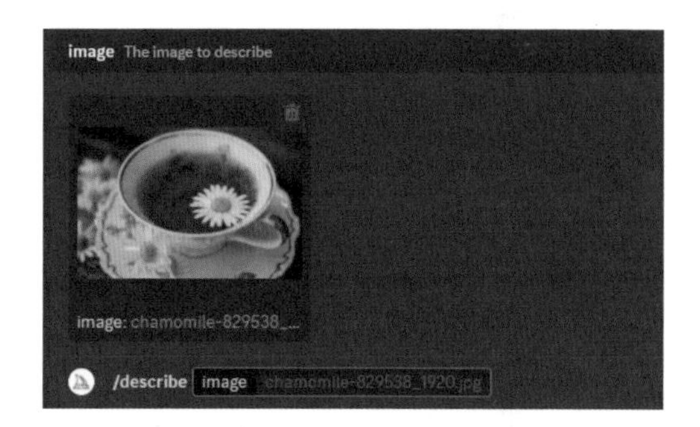

3. 分析 MJ 生成的提示关键词，在图片下方单击认可的某一组提示词的序号按钮。

4. 笔者在此单击的是 1 号按钮，并在打开的文本框中对提示词进行修改。

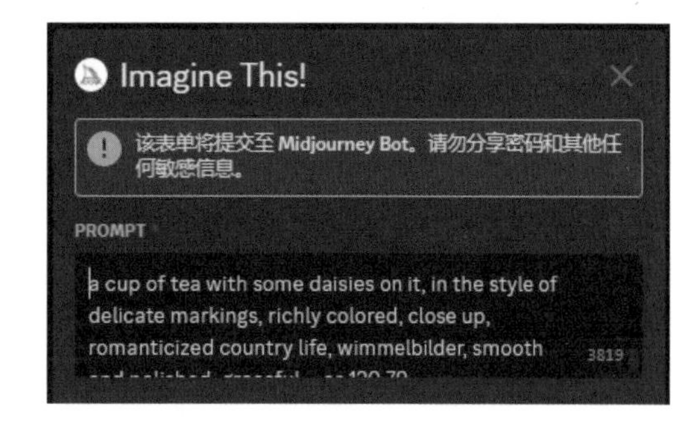

5. 第一次生成的效果如下图所示，可以看到效果并不理想，因此，需要调整提示词。

6. 分析提示词后发现，由于 MJ 分析生成的提示词中没有针对视角及杯子的描述词，因此笔者将提示词修改为：aerial view,a cup of tea with some daisies on it,victoria style ceramic tea cup, luxury , richly colored, close up, smooth and polished, graceful，并添加了参数 --ar 3:2 --v 5 --q 2 --s 100，最终得到以下效果图。可以看出，位于左下角的 3 号方案基本能够满足要求。

用 show 命令显示图像 ID

MJ 中生成的每一张图像都有一个唯一的 ID 值。

通过图像的 ID 值，可以重新在生成列表中显示这个图像，以便在此基础上获得图像的 seed 值，或对此图像执行衍变操作。

从文件名中获得 ID

如果已经下载了自己的图像，可以通过查看图像的文件名获得 ID。

例如，一个图像的文件为：llbb_a_dog_in_blue_suit_clothes_and_a_cat_in_red_suit_clothes_s_e7978f3c-b012-43ec-bd7e-4e16db111bd0.png，其中，e7978f3c-b012-43ec-bd7e-4e16db111bd0 为 ID 值。

从网址中获得 ID

如果在自己的作品页面打开了图像，则可以从网址栏中找到 ID。例如，笔者打开了自己的一张图像，地址栏会显示 https://www.midjourney.com/app/jobs/217cef0c-d8ed-44a6-9dfb-98633c2573e8/，其中，217cef0c-d8ed-44a6-9dfb-98633c2573e8 为 ID 值。

通过互动获得 ID

除了上述方法，还可以使用前面讲解过的获得 seed 数值的方法来获取 ID 值，seed 值上方的 Job ID 后面显示的就是 ID 值。

用 ID 重新显示图像

获得 ID 值后，可以使用 /show 命令重新显示此图像。

显示图像后，可以单击 u1 ~ u4 按钮来放大图像，或单击 v1 ~ v4 按钮来衍变图像。

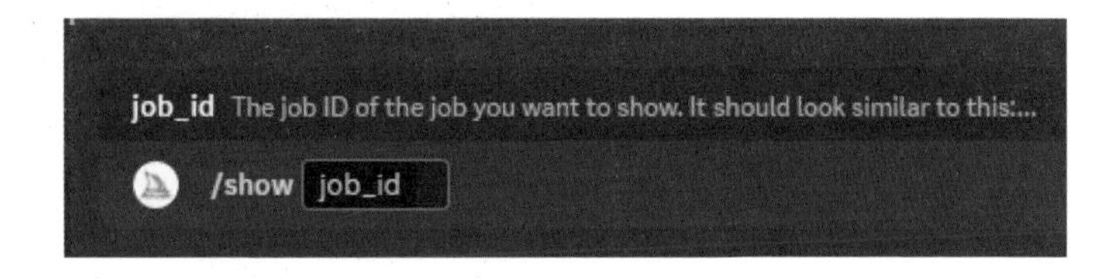

用 remix 命令微调图像

如前所述，当生成四格初始图像时，单击 v 按钮，可以在某一张初始图像的基础上，再执行衍变操作生成新的图像。此时执行的衍变操作基本上是随机的，创作者无法控制衍变的方向与幅度。

为了增加效果的可控性及图像的精确度，MJ 新增了 /prefer remix 命令。执行此命令可以进入 MJ 的可控衍变状态，MJ 将弹出如下提示，提示创作者进入了 remix 模式。

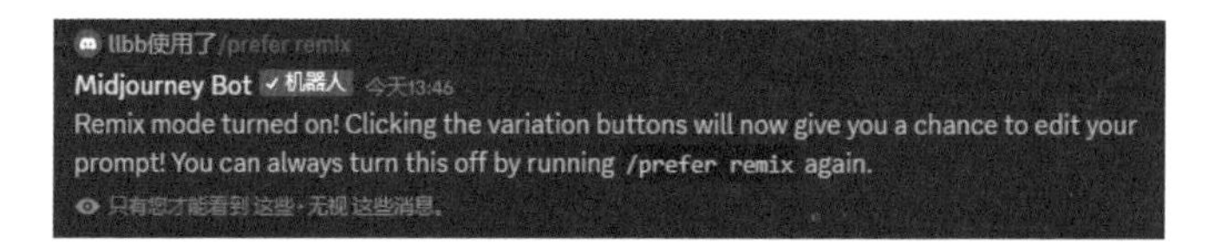

此时，再单击 v 按钮，将弹出一个提示语修改框，在此框中修改关键词后，即可使 MJ 在衍变时更精确，得到的效果也更可控。

例如，笔者使用提示语 wonderful ethereal ancient chinese white gold dragon floats over a crazy wave sea, high quality, cloudy --s 1000 --q 2 --v 5 --ar 3:2 生成了下面的图像，其中龙的身体被定义为金色。

针对右上角的图像，如果希望将龙的颜色修改为银色，则可以单击对应的 v2 按钮。

在弹出的 Remix Prompt 对话框中将 white gold dragon 修改为 silver dragon，便可以得到下面展示的银色龙身。

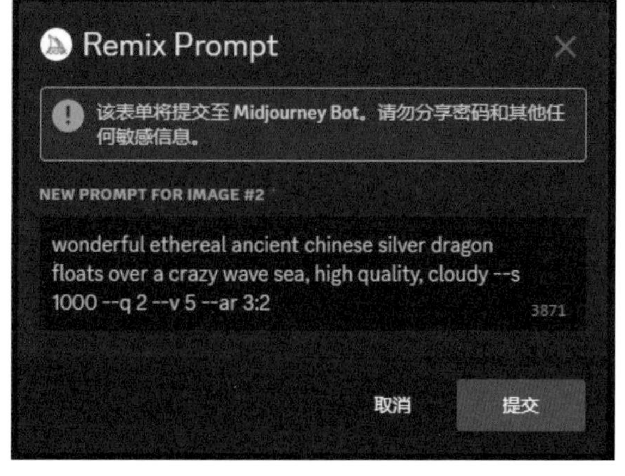

可以根据需要再次做衍变处理。例如，笔者单击 v1 按钮后，在弹出的对话框中添加了 red glowing eyes 关键词，为龙增加发红光的眼睛，此时得到如右侧所示的图像。

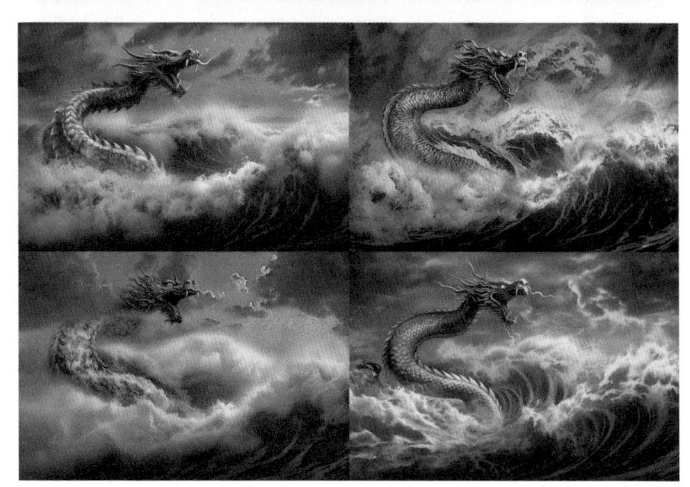

用 info 命令查看订阅及运行信息

在 MJ 命令区输入或找到 /info 命令，直接按回车键，可以显示如下信息，以查看自己账户的运行情况。

your info（你的信息）

subscription: standard (active monthly, renews next on 2023 年 4 月 27 日晚上 8 点 54 分）

订阅：标准版（已激活，下次续订时间为 2023 年 4 月 27 日晚上 8 点 54 分）

job mode: relaxed（任务模式：轻松模式）

visibility mode: public（可见性模式：公开）

fast time remaining: 0.85/15.0 hours (5.64%) [快速时间剩余：0.85/15.0 小时（5.64%）]

lifetime usage: 7043 images (118.97 hours) 已使用情况：7043 张图片（118.97 小时）

relaxed usage: 1575 images (24.77 hours) 轻松模式使用情况：1575 张图片（24.77 小时）

queued jobs (fast): 0 待处理的任务（快速）：0

queued jobs (relax): 0 待处理的任务（轻松）：0

用 shorten 命令对提示语进行分析

此命令可以帮创作者分析自己撰写的提示语，并通过删除线来指出哪些单词无效，哪些单词是关键作用。

操作时在 MJ 命令区输入或找到 /shorten，然后输入一段提示语，回车后，MJ 就会显示左下图的提示，其中 MJ 认为，加粗显示的提示词更重要，而有删除线的提示词被被认为无用。

如果点击下方的 show details 按钮，则会显示右下图的提示。在此，MJ 以数值以及阴影图的方式，对各个单词进行标注，例如，在下面的示例中，MJ 认为 frost 最重要。

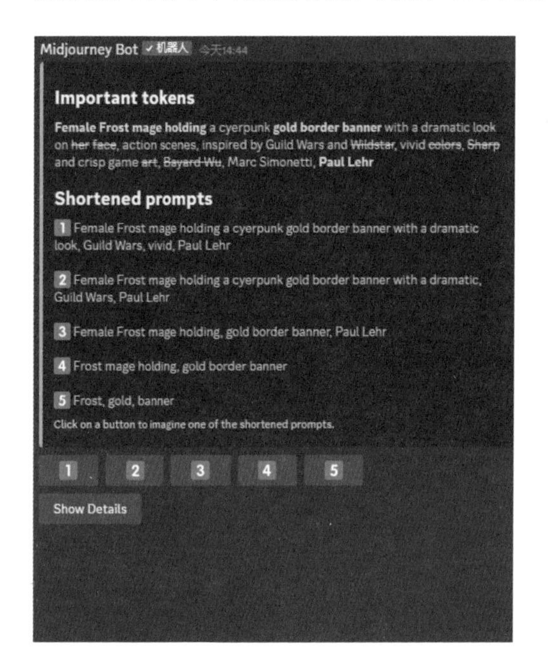

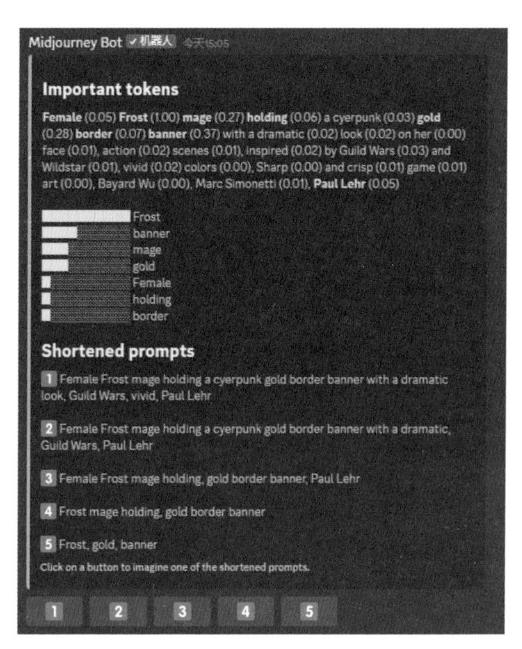

除了分析整段提示语，MJ 还会生成四条新的提示语条，分别点击下面的 1~4 号按钮，即可直接使用这些提示语生成图像。

不得不说的是，虽然这个功能的初衷是好的，但根据笔者的测试，此命令仅能够起到很小的参考意义。

因为 MJ 实际上并不能很好地理解创作者的构思，尤其是当创作者使用的是翻译软件通过直译的形式输入一段提示语时，本身很有可能存在表达不规范、产生歧义的情况，这就更容易使 MJ 产生误判。

因此，当创作者按 MJ 的建议使用给出的缩短后的提示语，有时会发现图像的整体效果与初始臃肿的提示语生成的效果有很大的区别。

用特定提示语获得一致性角色

由于 MJ 在生成图像时的随机性，使用常规方法很难获得同一角色，不同角度或场景的图像。但如果希望使用 MJ 生成有连续情节的漫画书或剧情系列套图，角色的一致性非常重要。

要解决这一问题，目前有两种方法。

第一种是使用一组特定的语句 split into 4 different images, shot from multiple angles，通过这组语句生成不同角度的图像。

另一种是使用换脸技术，将不同图像的角色，更换为同一张脸，以保持角色的一致性。

下面讲解第一种方法。

例如，以下面的描述语句 a boy surfing at the beach, blue waves, palm trees, 3d, unreal engine; split into 4 different images, shot from multiple angles --niji --style expressive --ar 3:2 --s 800，可以生成如下图所示的四格图像。

其中，第 2 幅（右上）、第 4 幅（右下）中人像的表情与动作都比较完整，因此可以按 u 按钮，生成大图，然后，将这些大图在 Photoshop 中进行裁剪，就可以得到同一角色在不同角度下的多幅图像。

值得一提的，在此基础上可以使用 zoomout 按钮扩展场景。左下图所示为扩展前的图像，右下图所示为扩展后的图像。

另外，即便这个提示词也有一定概率，无法获得令人满意的结果，其中包括生成的图像没有人像，如下图所示，或直接就没有生成分割成为几块的图像。

使用换脸机器人获得一致性角色

如前所述，要获得一致性角度还可以使用换脸机器人，基本步骤如下所述。

1. 在浏览器中打开链接https://discord.com/oauth2/authorize?client_id=1090660574196674713&permissions=274877945856&scope=bot，将换脸机器人邀请到自己的服务器中。

2. 在弹出的对话框中选择自己的服务器，如左下图，然后点击“继续”按钮，随后在新的对话框中点击“授权”按钮，此时就会弹出成功的对话框，如右下图。

3. 按上述过程操作后，在提示区输入 /，则可以看到换脸机器人及其命令，如下图所示。

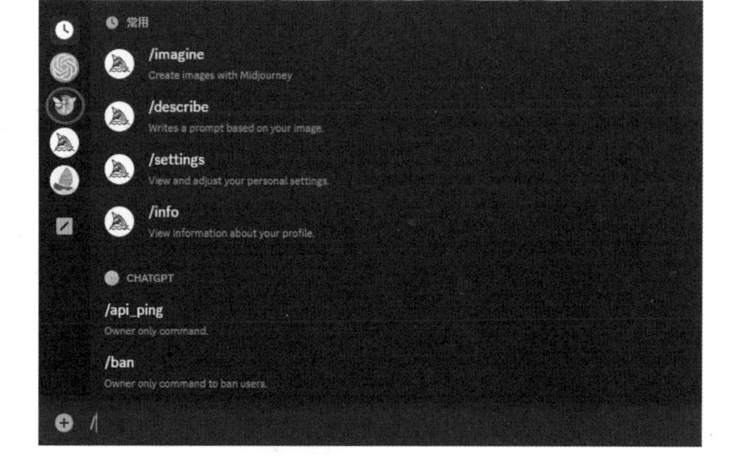

4. 使用 insightface 的一键换脸功能需要先定义一个 ID 名称，可以在命令行输入框中输入命令 /saveid，以给源头人脸图像命名，例如，在此笔者使用的是 ID 名称为 hb，然后将源头人脸图像，在此为奥黛丽赫本的图像拖入命令行窗口中，也可以点击窗口进行选择。要注意使用的人脸图像最好是正面、无遮挡、高清图像。

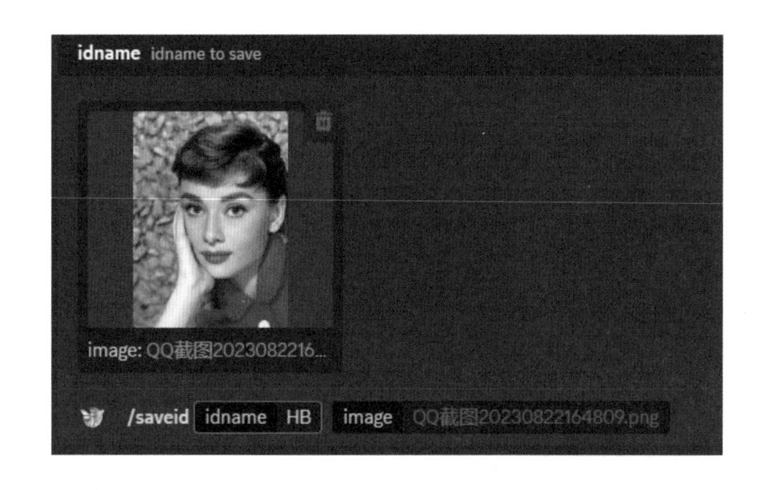

5. 按回车上传图像后，换脸机器人，将显示创建成功的提示，如右图所示。

6. 接下来是换图的步骤。在命令提示区输入命令 /swapid，在 idname 后面输入第 4 步定义的 ID，并上传要换脸的图像。

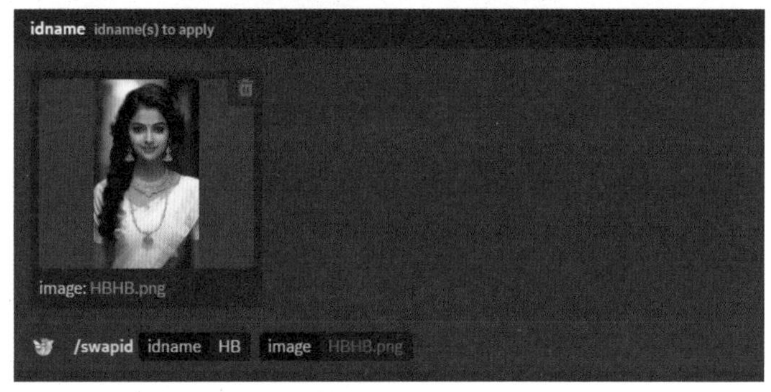

7. 回车后即可获得换脸后的图像。需要注意的是，一个账号每天只有 50 次换脸操作额度。

第 3 章 用 Photoshop AI 功能对 Midjourney 图像局部精调

Photoshop 新增 AI 功能简介

随着 AI 技术快速发展，许多软件和工具也不断适应并整合 AI 技术，不仅文字编写、PPT 制作实现了高效 AI 式生成。各类绘图软件也纷纷加入了相应的功能。

Photoshop 作为业界著名的图像编辑软件，自然也加入了 AI 绘画功能，进一步扩展了其功能和创作能力，为用户带来了更多创作的可能性。

通过使用 Photoshop 的 AI 技术，创作者不仅可以直接生成素材图，还可以将其用于去除杂物、修补图像、融合图像等方面，此外，还可以与 MJ 联合应用。

例如，左下图展示的是原图像，为了将照片修改为练习演讲主题的照片，使用 AI 功能去除右侧的女性，并增加花瓶与笔记本电脑，便会得到右下图所示的照片。

从整体效果上来看，这一修改几乎毫无破绽，无论是颜色还是光影都非常自然逼真。

又如，在右侧展示的两幅图中，左侧为使用 MJ 生成的原图，为了将这张图应用在较长的商品详情图中，使用 Photoshop 的 AI 技术进行了扩展，使图像显示了更多细节，图像的景别也从特定转变为中景。

下面将分别讲解如何在各种创作任务中，使用强大的 Photoshop AI 技术。

Photoshop AI 功能基本使用流程

PS AI 功能的使用方法非常简单，只需要简单两步即可完成。第一步为创建选区，第二步为在上下文任务栏的“创成式填充”按钮，并以英文输入任务关键字即可。

在此特别要强调的是，在最新的 PS 版本中，可以直接输入中文获得 AI 生成图像，但由于其机制是将中文翻译为英文，再根据英文生成图像，因此为了避免翻译误差，建议直接输入英文。

下面通过一个具体小案例来进行讲解。

1. 打开任意素材照片，在此笔者打开的是一幅飞行中的天鹅的特写照片，如下图所示。

2. 按 D 键，将背景色设置白色，在工具箱中选择裁剪工具 ![裁剪工具]。使用此工具，在照片中点击一下，然后向外拖动裁剪框句柄，以扩展画布，得到类似于如图所示的效果。

3. 在工具箱中选择魔棒工具 ![魔棒工具]，点击画面中的白色区域，将其选中。然后，点击“选择”—“修改”—“扩展”命令，在弹出的对话框中输入 10，使选区能够包括一部分原始图像，如下图所示。

4. 点击上下任务栏中的“创成式填充框”按钮，在输入框中输入 lake，再点击“生成”按钮，即可得到 AI 自动填充的图像，如下图所示。

在上面的这个小案例中，虽然笔者输入的提示词是 lake（湖），因此得到有湖面的图像。各位读者在实践练习操作时，可以根据自己练习的素材，输入其他提示词。

了解并应对 PS AI 的随机性

PS AI 在生成图像时有一定的随机性，这也意味着生成的图像有时好有时差，而且每次生成的效果不相同。

这是因为，当 PS AI 功能收到生成指令后，将基于 AI 算法从已经海量的图库图像中，依据提示词寻找符合要求的若干图像，然后通过 AI 算法，提取这些图像的特征，绘制出一张新的图像，并将其与当前工作图像进行融合。

在这个过程中，软件会随机抓取符合要求的图库图像，因此生成的图像质量参差不齐。

要选择质量更高的生成图像，可以按下面的方法操作。

1. 按 F7 键，显示“图层”面板，选择名称为“生成式图层”的图层。

2. 点击“窗口”—“属性”命令，显示“属性”面板。

3. 在“属性”面板中分别点击生成的三个方案，从中选择质量更高的图像，例如，在上一例中，另外两个方案分别如下所示。

4. 除上述方法外，还可点击“生成”按钮，生成新的方案，如左下图所示。

5. 也可以在“属性”面板的提示词输入框中输入新的提示词，以获得新的方案。例如，笔者添加了 sunset（落日）提示词后，获得如右下图所示的效果图。

了解 PS AI 的局限与不足

虽然，PS AI 有强大的图像生成功能，但由于此功能开发的时间并不长，因此目前仍然有许多不足之处，了解这些不足之处，有助于创作者在工作中扬长避短。

总体来说，PS AI 有较明显的两处不足。

精确度不足

左下图所示的为原图像，右下图展示的是将人像的背景选中，在提示词输入框中输入 beach（海滩）后得到的效果，可以看出，人物的面部、手、脚等处均出现了扭曲。

虚拟类题材效果不足

与 MJ 等 AI 绘图软件不同，PS AI 功能的训练库来源于 adobe 的图库，这些图库中绝大多数属于实拍类照片，这造成了当使用 PS AI 生成虚拟、幻想类题材图像时，效果欠佳。

例如，下面是笔者使用 robot 提示词生成的效果，完全无法与 MJ 生成的效果相提并论。

语义理解不足

由于 PS AI 生成图像依赖于提示词，因此提示词的理解程度将直接影响生成图像的效果。从目前来看，PS AI 只能理解简单的提示词，例如，PS AI 针法区别方位、数量、逻辑关系，因此，在运用提示词时要尽量使用简单的词与语句。

理解 PS AI 与选区间的三种关系

如果将提示词形容成为给 PS AI 的行动指令，那么创作者创建的选区就是给 PS AI 的行动方位，即选区会限定 PS AI 生成的图像的位置。

而且选区还会限定 PS AI 所生成的图像的大小。例如，左下图展示的是使用较大的选区生成的鸟的图像。右下图所示的为使用较小的选区生成的图像，大小对比一目了然。

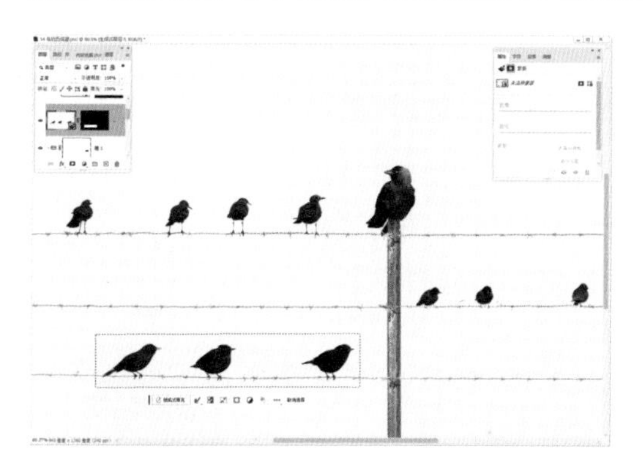

除了位置、大小外，选区的形状也会影响生成的图像。例如，左下图所示为绘制一个横向矩形选区，并在提示词输入框中输入 dog 后生成的图像。右下图所示为，按道路方向绘制一个梯形选区后生成的图像。很明显，如果希望小狗沿着道路行走，要使用梯形选区。

当图像中存在多个选区时，有时 PS AI 会根据这些选区生成指定的图像，但更多情况下，无法精确依据这些选区生成图像，但随着 PS AI 功能逐步完善，相信在更新迭代几个版本后，创作者可以使用多个选区精确地生成图像。

利用 PS AI 生成实拍素材照片

可以毫不夸张地说，掌握 PS AI 就相当于拥有了一个小型的素材库，因为使用 PS AI 可以生成高质量的实拍素材照片。只需要新建一个图像，然后按 Ctrl+A 键执行全选操作，再在提示词输入框中输入希望得到的素材的英文描述即可。

例如，左下图为笔者使用 grass, close up, sun, blue sky, wide angle 得到的素材，右下图为使用 overlooking a vast expanse of green grass, undulating and uneven, with a gentle breeze and a dynamic blurry effect, blue sky and cloudy skies 得到的有动感模糊效果的草原素材。

左下图提示词为 cream puff, fruit on the table, a glass of hot chocolate, hyper detailed photography, food photography, beautiful lighting, dark background。右下图提示词为 round cheese puff pastry, white background。

左下图提示词为 stock photo style, oranges fall into water, water splash, realistic photography,black background, close up。右下图提示词为 fantastic clouds, snowy mountains, sandy terrain, long exposure effects, award-winning black and white photos, master style photography。

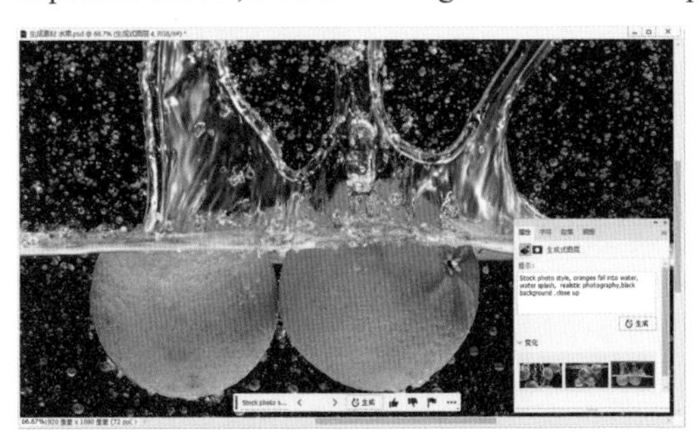

用 PS AI 修补图像的方法与技巧

虽然，ps 提供了“编辑”|“内容识别填充”命令、修补工具 、修复画笔工具 ，但在修补图像方面，均不如 PS AI 功能。

例如，左下图展示的原图，如果想在这个图像中移除手机与黑色的手机支架，使用 PS AI 功能可以得到中间所展示的效果；而如果使用使用同样的工作时间，使用“内容识别填充”命令、修补工具 、修复画笔工具 只能得到如右下图所示的效果。

使用 PS AI 修补此图像时，首先要使用套索工具 选中左下角的手机，选择时注意要将台面上的阴影也选中，如左下图所示。然后直接点击“创成式填充”按钮，再直接点击“生成”，即可得到右下图所示效果，可以看到，效果相当不错。

图像中其他的区域也按此方法操作，即可获得不错的效果。

如果在操作时遇到如右图所示的提示对话框，可以尝试在提示词输入框中输入 remove noise，再次执行生成操作。

用 PS AI 功能为图像添加细节

使用 PS AI 功能可以为图像添加逼真的细节，这对于增加图像的可信度、完善图像的构图很有帮助。

例如，左下图所示为原图像，在图像右下角绘制一个选区，并使用 multiple cobblestones 提示词，可以添加几个礁石，如下方中间图所示。在右上方绘制一个小矩形选区，并使用 boat 提示词，可以添加一艘船，如右下图所示。

在此基础上，于图像左上角绘制几个矩形选区，并使用 bird 提示词，则可以添加飞鸟，如下图所示。

在下面的示例中，笔者使用 water splashes caused by objects rushing out of the water surface 提示词增加了水面水花。

在下面的示例中，笔者使用 a tree with lush branches and leaves 及 the tall and steep k2 peak in shadow under star sky 提示词，优化了不好看的树与山峰，如下图所示。

利用 PS AI 处理过曝的照片

在摄影中过曝是很常见的照片问题，通常在光线过强、被拍摄对象颜色较浅、曝光参数不恰当的情况下，均容易出现照片过曝光的问题，此时照片会出现大面积白色，如左下图所示。

针对这类问题，可以使用 PS AI 功能来有效解决，得到如右下图所示的效果图。

1. 打开过曝的图像后，点击“选择”—“色彩范围”命令，在弹出的对话框中，用吸管点击图像中的白色区域，如图所示。此时对话框下方有一个黑白图像，白色区域代表选中的过曝区域，如果感觉区域较小，可以拖动对话框中的“颜色容差”滑块，以扩大选取范围。

2. 点击“确定”按钮，退出对话框，即可得到如右下图所示的选择区域。

3. 直接点击“创成式填充”按钮，再直接点击“生成”，即可得到过曝区域被修复后的效果。

按同样的原理，可以针对欠曝的图像进行处理，只是在使用“色彩范围”命令时，要使用吸管工具点击图像中的黑色区域。

使用 PS AI 处理人像照片

在拍摄人像照片时，受客观条件限制，可能会由于各种客观条件导致照片有这样或那样的小瑕疵，此时都可以考虑使用 PS AI 功能进行修改处理。

例如，在左下图所示为原片，右下图所示为使用 PS AI 处理后的照片。对比前后效果可以发现，笔者使用 PS AI 修补了过曝光的白色衣服、添加了白色珍珠耳坠、更换了珍珠项链，给模特手中添加了小花、稍微修改了发型，以上操作有些没有使用提示词，部分使用了非常简单的提示词，如 pearl necklace、holding a flower、slightly curled long hair, with a fluffy and glossy feel。

操作时要注意控制选区的大小、形状，只要填写正确的提示词，就能获得不错的效果，这个案例充分证明了 PS AI 在人像照片修饰方面的潜力。

使用 PS AI 为人像换装

除了对人像照片进行小修小改外，还可以使用 PS AI 对人像的衣服进行大面积替换，例如，在下面展示的四张照片中，除左图为原图外，其他三张图片均为使用 PS AI 进行换装后的图像，从效果来看，非常自然、逼真。

要执行这样的换装操作，除了在创建选区时，尽可能将要更换的衣服全部选中外，还要注意修改提示词，并综合使用修补工具 、修复画笔工具 进行修补操作，因为在执行大面积换装操作时，有可能在图像的局部会出现明显的瑕疵。

例如，在下面展示的三张图分别在衣领、脖子、手部均出现了明显的瑕疵，此时，需要创作者具有综合使用 PS 各种功能与工具的能力，去除这些瑕疵。

使用 PS AI 处理绘画类作品

如前所述，虽然在现阶段，针对虚拟类、绘画类题材，使用 PS AI 无法获得较好的生成图像，但 PS AI 在处理绘画类作品时，仍然表现出了较高的水准。

换言之，只要原素材图像的信息量足够多、足够全，PS AI 功能就可以根据已有的数据信息推算出要处理的局部或可生成新图像。

例如，下面展示的分别是对抽象画风的绘画作品、写实油画、中国画处理前后的效果，可以看出，使用 PS AI 功能扩展出来的新区域，很好地维持了原画作的风格。

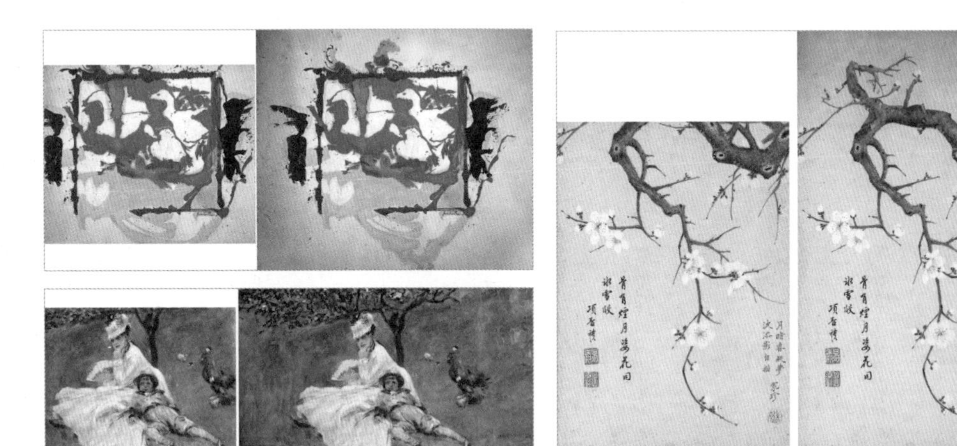

即便处理的是虚拟的主题或插画，效果也相当好，这也正是笔者在本章前面提到的一个观点，在掌握 Midjourney 后，依然是一定要掌握 PS AI 功能，以更加精确地控制图像的局部。

第 4 章 Midjourney 写实摄影照片创作 76 例

生成式摄影照片的常规应用

随着技术不断更新迭代，目前已经能够使用 MJ 生成真假难辨的照片质量图像。对于媒体或与视觉设计相关的工作者来说，可以根据文章或设计主题很方便地生成自己需要的图像，而且还没有版权的顾虑与成本的压力。此外，有许多特定的图像根本无法拍摄，或很难从图片库中找到，因此也可以借助 MJ 来生成。

总之， 掌握使用 MJ 生成摄影质量照片的技术，有助于每一位工作于视觉影像相关领域的创作者以更低的成本、更高的效率，获得自己所需要的一切素材图像。

生成摄影照片常用关键词

景别关键词

特写 close-up、中特写 medium close-up、中景 medium shot、中远景 medium long shot、远景 long shot、全身照 full length shot、大特写 detail shot、腰部以上 waist shot、膝盖以上 knee shot、面部特写 face shot。

光线关键词

顺光 front lighting、侧光 side lighting、逆光 back lighting、侧逆光 rim lighting、体积光 volumetric lighting、工作室灯光 studio lighting、自然光 natural light、日光 day light、夜光 night light、月光 moon light、丁达尔光 god rays。

视角关键词

前视 front view、侧视 side view、后视 back view、低角度仰视视角 low angle shot、俯拍视角 overhead、常规视角 eye-level、顶视图 top view、广角视角 wide angle view、全景视角 panoramic view、鸟瞰视角 aerial view、卫星视角 satellite view。

描述姿势与动作的关键词

站立 stand、坐 sit 、躺 lie、弯腰 bend、抓住 grab、推 push、拉 pull、走 walk、跑步 run、跳 jump、踢 kick、爬 climb、滑行 slide、旋转 spin、拍手 clap、挥手 wave、跳舞 dance、握拳 clenched fist、举手 raise hand、敬礼 salute、动感姿势 dynamic poses、功夫姿势 kung fu poses。

描述面貌特点的关键词

眼睛 eyes、眉毛 eyebrows、睫毛 eyelashes、鼻子 nose、嘴巴 mouth、牙齿 teeth、嘴唇 lips、脸颊 cheeks、下巴 chin、额头 forehead、耳朵 ears、颈部 neck、肤色 skin color、皱纹

wrinkles、胡子 beard/mustache、头发 hair 等关键词，对面部重点部位进行描述。

描述表情、情绪的关键词

高兴 happy、微笑 smile、平静 calm、惊讶 surprised、愤怒 angry、悲伤 sad、困惑 confused。

描述年龄的关键词

婴儿 infant、幼儿 toddler、小学生 elementary schooler、中学生 middle schooler、青年 young adult、中年人 middle-aged、老年人 elderly/senior/old man/old woman，也可以使用具体的数字来描述年龄，如四十五岁 forty-five years old、五十五岁 fifty-five years old。

描述服装的关键词

休闲风格 casual style、运动风格 sporty style、田园风格 rural style、海滩风格 beach style、优雅风格 elegant style、时尚潮流风格 fashionable style、正装风格 formal style、复古风格 vintage style、文艺风格 artistic style、简约风格 minimalist style、摩登风格 modern style、民族风格 ethnic style、花式风格 fancy style、波希米亚风格服装 bohemian style 、洛丽塔风格服装 lolita style、牛仔风 cowboy style、工装风 workwear style、汉服风格 hanfu style、维多利亚风格 victorian style。

生成风光照片常用关键词

山脉 mountain range、山峰 peak、峡谷 canyon、悬崖 cliff、河流 river、瀑布 waterfall、湖泊 lake、海滩 beach、海岸 coast、半岛 peninsula、岛屿 island、草原 prairie、沙漠 desert、高原 plateau、丘陵 hill、森林 forest、草地 meadow、湿地 wetland、火山 volcano、冰川 glacier、峡湾 bay、峁地 terraced landscape、沙丘 dune、花海 flower fields、石林 stone forest。

生成美食照片常用关键词

比萨 pizza、汉堡 hamburger、烤鸡 roast chicken、烤牛肉 roast beef、意大利面 spaghetti、牛排 steak、蘑菇汤 mushroom soup、沙拉 salad、烤三文鱼 grilled salmon、面包 bread、烤鸭 roast duck、豆浆油条 soy milk and fried dough sticks、红烧肉 braised pork belly、炒饭 fried rice、烤串 grilled skewers、炸鸡 fried chicken、饺子 dumplings、馄饨 wonton 等。

生成花卉照片常用关键词

玫瑰 rose、菊花 chrysanthemum、牡丹 peony、梅花 plum blossom、樱花 cherry blossom、芙蓉 hibiscus、雏菊 daisy、郁金香 tulip、向日葵 sunflower、蝴蝶兰 butterfly orchid、康乃馨 carnation 紫罗兰 violet、月季 rose、满天星 baby's breath、矢车菊 cornflower、花瓣 petal、桃花 peach blossom、花蕊 pistil、花茎 flower stem、花圃 flower bed。

写实摄影照片 76 例

动物写实效果照片

ultrasharp photography, by the national geographic award winning nature, closeup portrait of wild wolf, side view, black background, --s 1000 --ar 2:3 --c 5 --v 5.1

by the national geographic award winning , low depth of field macro supermacro photography, bokeh effect ,closeup hippocampus japonicus --s 750 --v 5.1 --c 5 --ar 2:3

a group of flying ducks, in the style of dark gray and light brown, --v 5 --ar 3:2

a satisfied tiger on a log in the jungle --ar 3:2 --v 5

食品素材效果照片

three fresh tomatoes ,white background. sparkling water droplets reflect light, light reflection --ar 16:9 --v 5

sliced whole grain bread. on white background --v 5.1

close-up of cut different fruits fills the entire space, still life photography --ar 3:2 --v 5

branch fresh orange tree fruits green leaves with wager drops --ar 3:2 --s 250 --v 5.2

食材创意效果照片

https://s.mj.run/63g0znpigcg young beautiful woman head and face made of vegetables, photography,vegetables background art by giuseppe arcimboldo --ar 3:2 --v 5

broccoli as bomb explosion,crack --ar 3:2 --s 800 --q 2 --v 4

食品写实效果照片

a chinese person made an appointment for a big breakfast in the breakfast shop, and the table was full of all kinds of delicacies, plane symmetry, graphics, 8k, high resolution

sichuan hotpot in china is filled with appetite and looks very mouth-watering. it contains vegetables, meat, ham, and many other foods cooked in a hotpot. the steam rises from it, and the bird's-eye view makes it very detailed and textured.

watermelon popsicles with ice on table. cool color,summer background --s 250 --style raw --ar 3:2 --v 5.2

a pitcher of sweet tea with a slice of lemon inside standing on a wooden table in the garden --s 250 --style raw --ar 3:2 --v 5.2

delicious fries, floating in the air in a plate, cinematic, food professional photography, studio lightning, studio background, advertising photography,

a big hamburger with cheese splashing through the air, in the style of vray tracing, fine art photography, intensity, organic and fluid --ar 2:3 --v 5.2 --s 750

风光写实效果照片

a winding road leads to the dense forest, with rays of light shining through the leaves onto the road, creating the "god rays" effect. volumetric lighting --ar 9:16 --v 5 --v 5

an aerial view photograph,mountain valley in the style of sebastiao salgado ,vivid color,dramatic backlight ,dramatic scene --ar 9:16 --v 5.2 --s 750

a photography of endless mountain ranges with abundant landscape and minimal sky, telephoto lens, sun light, minimalist composition --ar 3:2 --v 5.2

a beautiful photorealistic seascape at sunset shot with a long exposure of about 1 minute --v 5.2 --ar 3:2 --s 750

there are several small boats with fishing lights on the quiet lake, the vast lake, and the bustling city on the opposite side of the lake, gold hour, super wide angle --ar 16:9 --s 300 --v 5

there are several small boats with fishing lights on the quiet lake, the vast lake, and the bustling city on the opposite side of the lake,blue hour, super wide angle --ar 16:9 --s 300 --v 5

星轨及极光照片

panoramic,the streets of the city of schwaz from mountain top, moon light,long exposure, aurora background --ar 16:9 --v 5.1

panoramic,the streets of the city of schwaz from mountain top, moon light,long exposure,startrail --ar 16:9 --v 5.1

鸟瞰及无人机视角照片

busy city,day light,sunshine,satellite view --ar 16:9 --v 5

busy city,day light,sunshine,aerial view --ar 16:9 --v 5

chinese landscape,mountains, clear rivers, birds, clouds and mist, drone photography,majesticultra-vivid colors, --ar 3:2 --v 5

complex overpasses in forest, autumn, drone photography --ar 3:2 --v 5

极暗调效果照片

a silhouette image of a car,black background, rim light on car edge,side view --v 5 --ar 3:2

a silhouette image of a dog's head , black background, rim light, --v 5 --ar 3:2

微距效果照片

beautiful woman eyes,macro photo --v 5 --ar 3:2

bee,macro photo, flower background --ar 16:9 --v 5

城市玻璃幕墙效果照片

low angle image of typical contemporary office towers, tall structures with glass facades. financial and economic foundation concepts --ar 3:2 --s 250

look up angle,super wide angle,perspective low angle view, clean luxury modern buildings ,blue sky background, sun shine --s 250 --style raw --ar 3:2 --v 5.2

移轴效果照片

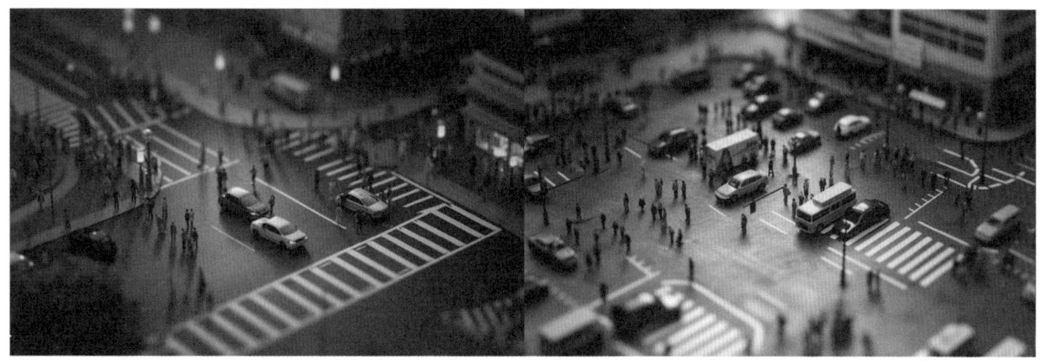

people are crossing the traffic light of a crossing road in the city, tilt-shift, aerial view, --ar 3:2 --v 5

动感模糊效果照片

lion running,motion blur, blur background,side view --v 5.1 --ar 3:2

a beautiful lady dressed in gorgeous chinese hanfu is dancing in an ancient chinese courtyard, motion blur, focus face,side view --v 5.1 --ar 3:2

红外效果照片

infrared photography,beijing street --ar 2:3 --s 500 --v 5

infrared photography,chinese landscape,mountains, clear rivers --ar 2:3 --s 500 --v 5

双重曝光效果照片

double exposure, portrait of an working man working in industry, --ar 3:2 --v 5

double exposure, portrait of a dancer in flower --ar 3:2 --v 5

柔美花卉摄影效果照片

blurry abstract blurry warm background with light highlights in delicate shades and blossom --upbeta --s 750 --v 5.1

a pink branch of a blooming apple tree in spring, photography style, shallow depth of field --v 5.1 --s 750

光绘效果照片

long skirts dancing girl, many red glowing lines and blue glowing lines, light painting,slow shutter speed,dark background ,wide angle,full portrait --ar 2:3 --v 5

a man rotates a spark-spraying fireball in a circle,light painting,slow shutter speed,motion blur,dark background ,no test,wide angle,full portrait --ar 2:3 --v 5 --q 2 --s 800

黑白效果照片

underground passage, volume light, man standing on stairs up，photography in the style of fan ho --ar 2:3 --v 5.1 --s 750

hongkong street , photography in the style of fan ho --ar 2:3 --v 5.1 --s 750

huangshan and huangshan pine trees shrouded in mist, black and white photos, award-winning photos, master works, --ar 2:3 --s 500 --v 5

big iceberg in ocean, dramatic backlight ,in the style of sebastiao salgado.dramatic scene --ar 3:2 --v 5.1 --s 750

云山雾水效果照片

chinese landscape,mountains, clear rivers, birds, clouds and mist, drone photography,majesticultra-vivid colors

散景效果照片

grape hyacinth flowers, beautiful bokeh, colorful flowers, against the light, --ar 3:2 --s 500 --v 5

masked woman washing table with gloves on, cleaning spray and sponge in hand, light interesting background, bokeh, effect lighting, --s 250 --style raw

色彩焦点效果照片

selective red color photography of city street, monochrome street scene, selective color effect, crowed,daylight --v 5 --ar 3:2 --s 800

selective yellow color photography , monochrome scene, selective color effect,a silver armor warrior facing a huge gold chinese dragon,burning fire, low angle,smoke,fire,rain,temple background --ar 3:2 --v 5

X 光透视效果照片

x-ray film of a the most beautiful flower blooming --s 250 --style raw --ar 3:2 --v 5.2

x-ray film of beijing city --s 250 --style raw --ar 3:2 --v 5.2

智慧城市效果照片

digital enterprise; realistic, technology based, connected, telecommunications --ar 3:2

network systems and iot city --s 550 --ar 3:2 --v 5.2

smart city, dot and gradient line connect city building, digital style, big data city technology concept --s 550 --ar 3:2 --v 5.2

retail boulevard with led screens in a trade fair, minimal, realistic, full of details, hyperreal, --s 550 --ar 3:2 --v 5.2

静物摄影效果照片

still life photography, elegant white plum blossoms in a vase, vibrant and radiant colors , beautiful lighting , award-winning photography by joanna kustra ,black background, --ar 3:2 --v 5.1 --s 100

still life in the style of carovaggio with fruit ,diffused lighting to create a painterly and romantic ambiance. --v 5.2 --ar 3:2 --s 750

画意效果人像照片

chinese woman in the bamboo forest, dressed in white translucent cheongsam , thin mist，full body,elegant，minimalism， in the style of soft focus romanticism, fashion bazaar style,back view,wind --ar 2:3 --v 5.2 --c 5

chinese female holding a lotus , elegant, big lotus in soft fog, hanfu chinese style, front view,full portrait,full body --ar 2:3 --v 5.2 --s 750

合影类人像照片

a photography of a diverse group of people from different ethnic backgrounds, genders, and ages, smiles --ar 3:2 --v 5.2 --s 750

happy employees putting their hands together, teamwork --ar 3:2 --v 5.2 --s 350 --style raw

幻想类人像照片

fairy queen wearing blue and white with gold horns, cherry blossom --ar 3:2 --s 750

展示用白模类人像照片

a mockup showcasing a white t-shirt being worn by a asian woman,beautiful bokeh --ar 3:2 --v 5.2 --s 750

创意时装人像效果照片

chinese supermodel fashion show with chinese kirigami style as the me full body composition , haute couture fashion show --ar 3:2 --v 5.2 --s 900

blue and white porcelain as the theme to shoot a chinese supermodel fashion show, blue and white, the background is a blue and white porcelain mural --ar 3:2 --v 5.2

汉服人像效果照片

full portrait,a beautiful lady dressed in gorgeous chinese hanfu is dancing in an ancient chinese courtyard, intense wide-range radial blur effect, focus face --v 5.1 --ar 10:13 --q 3 --s 500

full portrait,panoramic view,in a large garden outside a small cabin, there is a 35-year-old beauty with shoulder-length black hair, a calm expression, big eyes, and curved eyelashes. she is wearing a gorgeous hanfu , squatting in the flower bushes holding the flowers. two butterflies. wide angle,bokeh,back lighting --ar 2:3 --q 3 --v 5

极简人像效果照片

dynamic pose of goddess of glowing sun , in the style of japanese folklore-inspired art, inventive character designs, photo taken with ektachrome, zen minimalism --ar 2:3 --v 5.1 --s 650

dynamic pose of goddess of glowing sun , in the style of japanese folklore-inspired art, inventive character designs, photo taken with ektachrome, zen minimalism --ar 2:3 --v 5.1 --s 650

第 5 章 Midjourney 抽象纹理设计 28 例

抽象纹理素材在设计中的应用

使用 MJ 可以轻松生成海量抽象纹理素材，这些素材不仅可以上传到各图库网站中为创作者获取收益，也能广泛应用于自己的设计作品中。

下面是的四张截图分别是笔者在包图网、veer 图库、视觉中国图库、摄图网四个网站中以“背景”进行搜索获得的部分纹理。在使用 MJ 创作类似纹理时，不仅可以参考这些纹理图素材的颜色、构图，还可以针对某些网站提供的热门下载排行，仅生成稍微对路的素材。

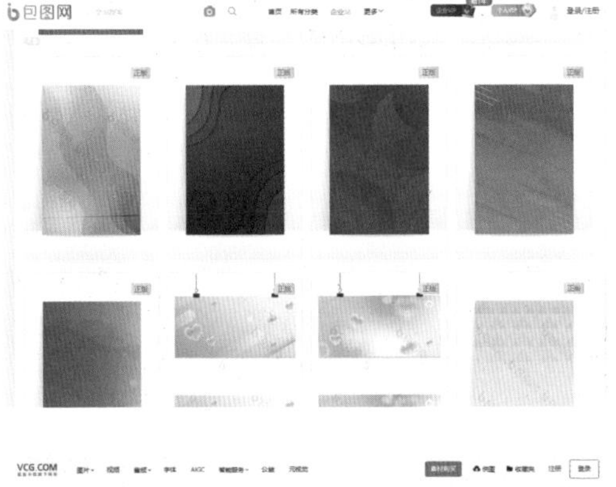

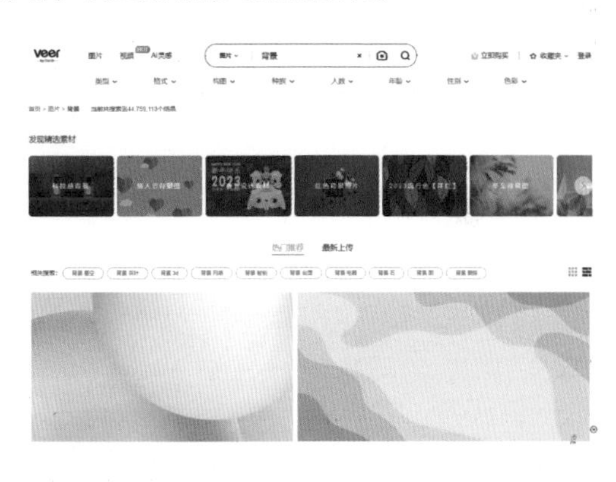

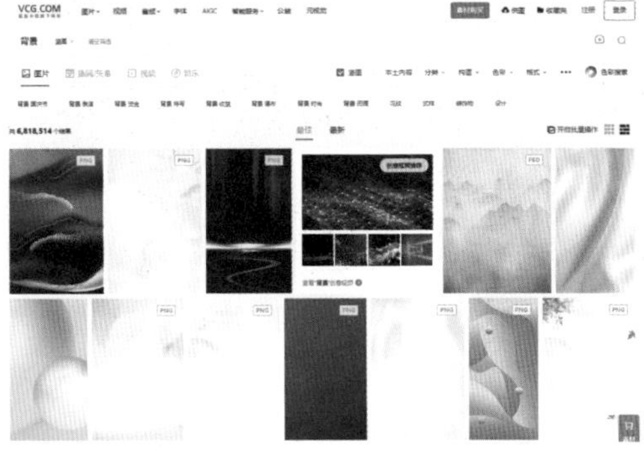

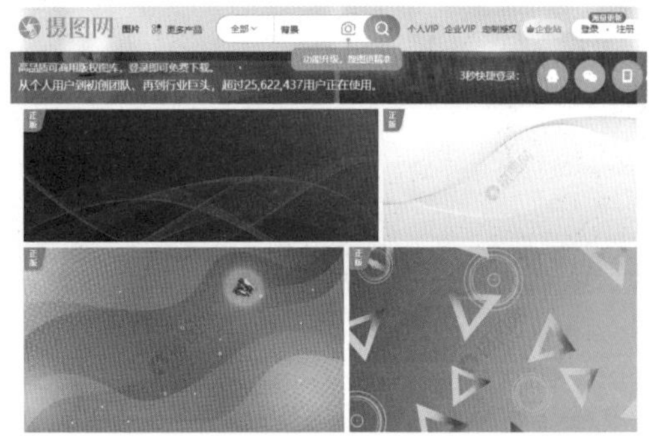

这些纹理素材可以广泛应用于广告、详情图、主图，甚至充当视频背景图，例如，右侧所示的课程主图背景使用的就是使用 MJ 生成的抽象背景纹理图。

抽象纹理素材效果 28 例

下面展示使用 MJ 生成的 18 种不同效果的纹理素材。掌握这些纹理所用提示语后，可以举一反三，创作出更多种类的纹理素材。

abstract light loop background , flashing lights and moving circles vector illustrator,emphasis on linear perspective, outrun, dark blue and orange, free-flowing lines, --s 750 --ar 2:3 --v 5.1 --v 5.1 --s 750

blue and gold abstract background image on dark black screen, chromatic sculptural slabs --s 750 --ar 2:3 --v 5.1

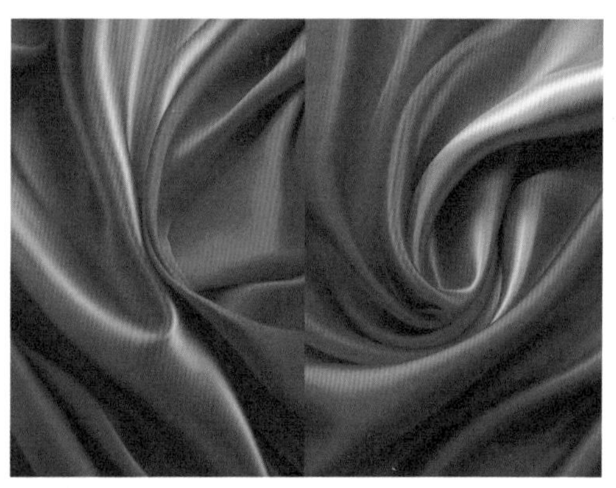

red and gold silk satin. color gradient. soft folds. shiny fabric. --s 750 --ar 2:3 --v 5.1

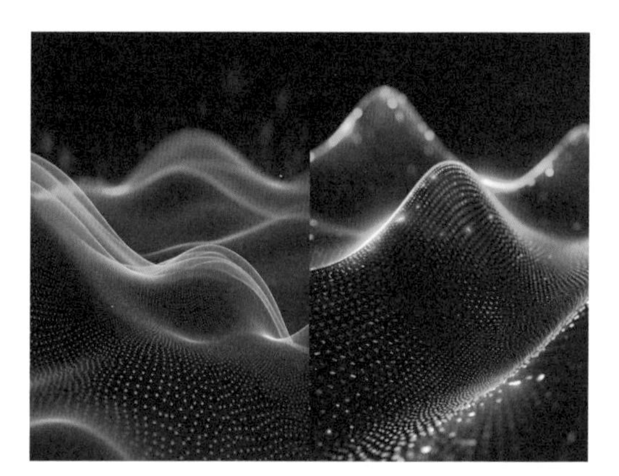

blue glowing waves,in the style of dotted,3d space,abstract blue lights,streamlined design,rhythmic lines, lens flare --s 750 --ar 2:3 --v 5.1

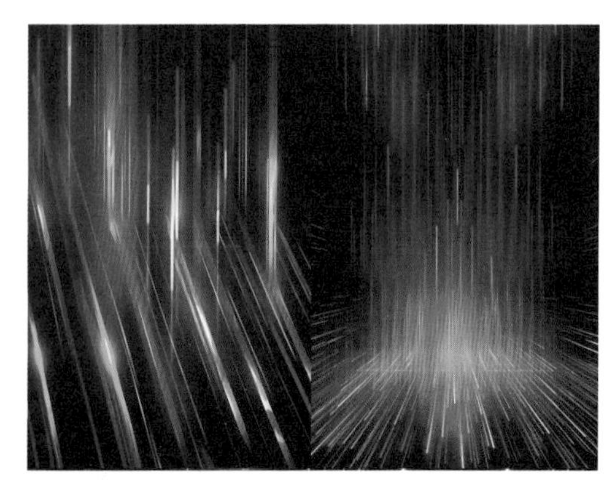

abstract neon background. colorful glowing lines. digital data transfer. futuristic style --s 750 --ar 2:3 --v 5.1

abstract bokeh background. light, christmas, bokeh, blur, light, gold, holiday, bright, christmas, yellow, colours, defocused, glow, decoration, orange --ar 3:2

a gold abstract flat background with glod dots, in the style of fluid and flowing lines, rollerwave, bokeh, hyper-realistic details --s 750 --ar 2:3 --v 5.1

geometric painting of the solar system with stars, --ar 2:3 --q 2

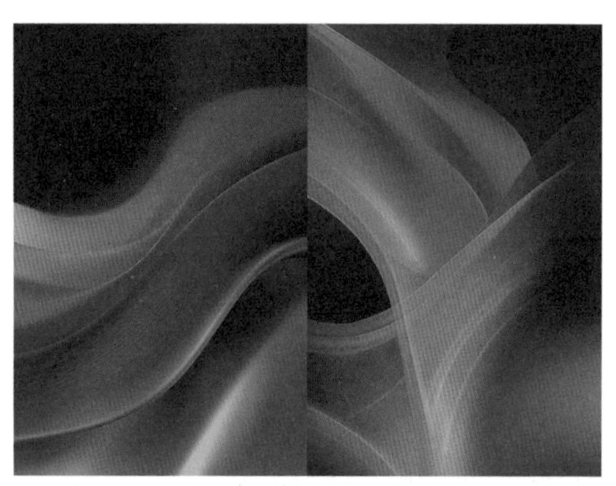

technology background with futuristic glowing lines, light indigo and red --s 750 --ar 2:3 --v 5.1

wave line,chinese ancient wind,the texture is rich and smooth, fine line wind, space feeling,more detail --ar 2:3 --s 750 --v 5.2

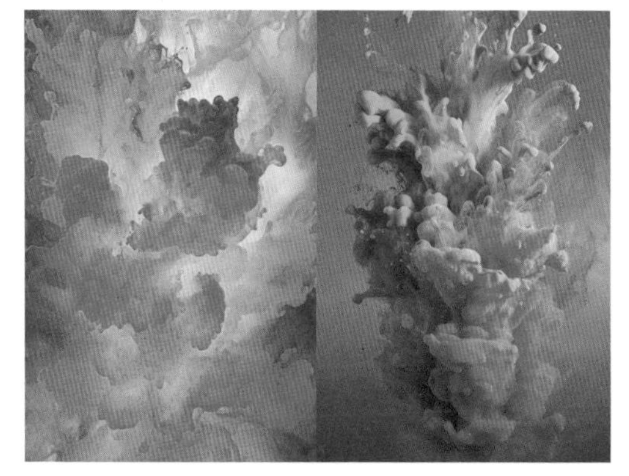

a neon color splash, in style of watercolour painting --s 750 --ar 2:3 --v 5.1

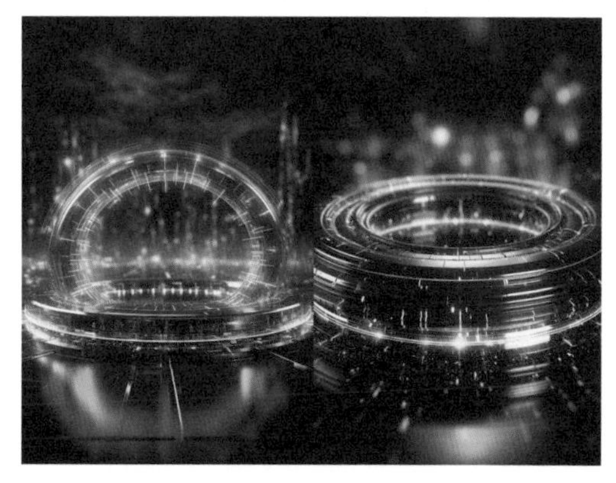

abstract futuristic circle core with glowing neon on dark blue background , moving high speed wave lines and bokeh lights --ar 2:3 --s 750 --v 5.2

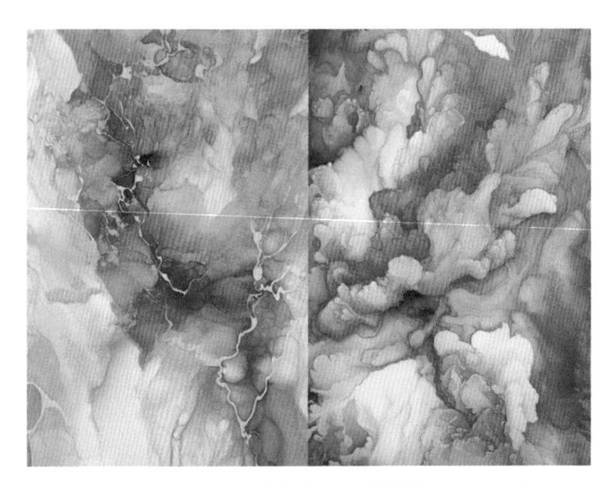

a colorful painting with lots of different colors, a watercolor painting by leticia gillett, abstract art, pink white, alcohol ink art, vibrant, ink block painting, --s 750 --v 5.1 --c 5 --ar 2:3

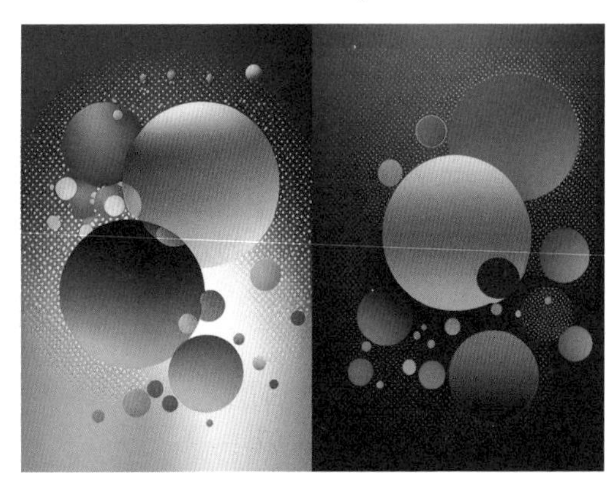

abstract, gradient gold circle and dot and blue soft colorful background. modern horizontal and gradient design --ar 2:3 --v 4 --v 4

minimal, glowing stroke, backlight abstract, simple, 3d, circle, rectangle layers colorful, unique design --ar 9:16

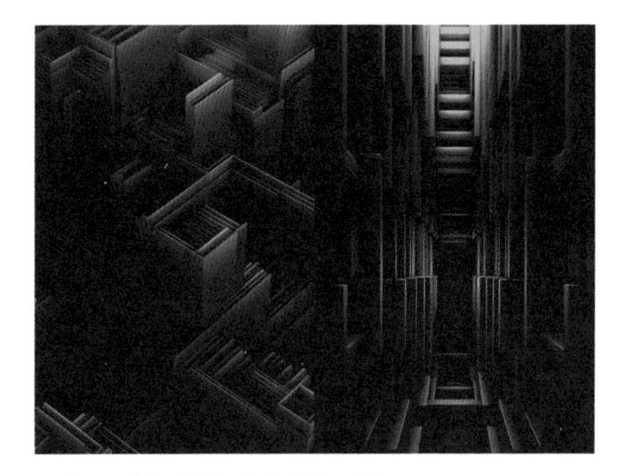

random minimalistic dark abstract 8k

random abstract minimalistic dark and gold metal regularly arranged rivets, glitter --ar 2:3 --v 5.2 --s 750

random minimalistic dark and gold metal abstract --ar 2:3 --v 5.2 --s 750

generate an abstract geometric pattern featuring circles. the design should display a harmonious balance and flow. the colors used should complement each other well, creating a visually pleasing pattern. the pattern may be repetitive or random but should have an overall sense of unity and balance. --ar 1:1 --tile --v 5

a seamless pattern inspired by the art of origami representing stars in a color palette of orange. the design should be clean, modern, and minimalistic, perfectly capturing the essence of paper folding art in a unique and stylish way. --v 5 --stylize 1000 --tile

cream on light blue damask wallpaper, seamless tiled, half drop, embossed, --tile --v5

cute cat,flat wallpaper pattern,pink --tile --v 5

an intricate pattern of pastel-colored bismuth,geometric iridescent shine,dominating blue shine, seamless design,soft light --uplight --q 2 --s 100 --tile --ar 2:3

abstract light loop background,flashing lights and moving circles vector illustrator,emphasis on linear perspective, outrun, dark blue and orange, free-flowing lines --s 750 --ar 2:3 --v 5.1

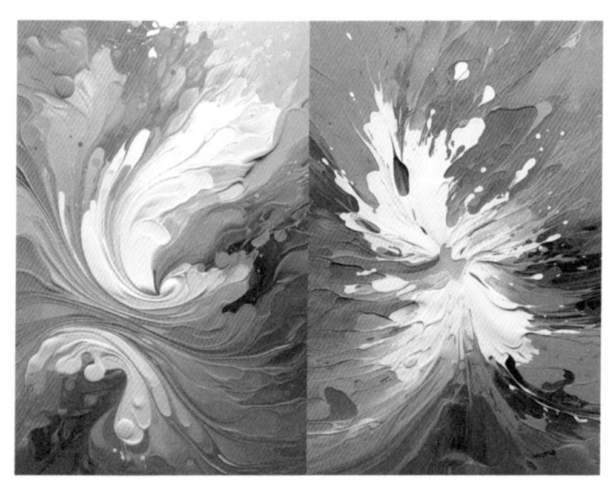

splashes of colorful paint in a cool and abstract design. big,space,pop --ar 2:3 --c 14 --style raw --v 5.2 --s 750

abstract watercolor art, central composition with blue and orange colors. fluidity and organic movement, harmony and balance --ar 2:3 --c 14 --v 5.2 --s 750

an abstract oil painting inspired by the bustling energy of city life, employing dynamic lines, intersecting shapes, bold color --ar 2:3 --c 14 --style raw --v 5.2 --s 750

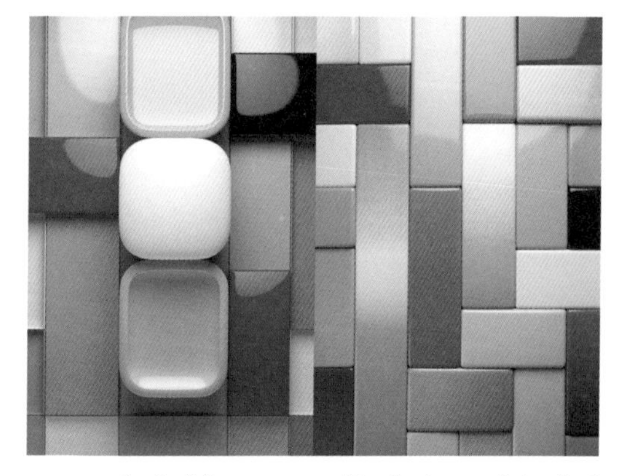

a group of colorful squares on a blue background, inspired by fernando gerassi, orange ray, orange racing stripes, vintage retro --ar 2:3 --c 14 --style raw --v 5.2 --s 750

第 6 章 Midjourney 箱包设计 17 例

常用箱包关键词

不同类型包的名称

复古帆布背包 vintage canvas backpack rucksack、男士真皮斜挎肩包 men's genuine leather crossbody shoulder bag、女士尼龙防水旅行背包 women's nylon water resistant travel backpack、中性户外运动徒步背包 unisex outdoor sports hiking backpack、时尚 pu 皮革笔记本电脑背包 stylish pu leather laptop backpack、休闲帆布肩包信使包 casual canvas shoulder bag messenger bag、轻便可折叠背包 lightweight foldable packable backpack、波希米亚流苏背包钱包 bohemian fringe tassel backpack purse、复古民族刺绣帆布背包 retro ethnic embroidery canvas backpack、女士高端多功能软质 pu 皮革手提包双层大容量背包 women high-end multifunction soft pu leather handbag double layer large capacity backpack、复古休闲帆布双肩背包手提包女士男士通用 brenice vintage casual canvas backpack handbag for women men、女士尼龙休闲防水肩包旅行妈咪包 women nylon leisure waterproof shoulder bag travel mummy bag、女士三件套流苏手提包斜挎包 women three-piece set tassel handbag crossbody bag、复古商务多口袋多功能防水可穿戴多用途背包公文包 ekphero vintage business multi-pockets multifunction waterproof wearable multi-carry backpack briefcase - ekphero 、男士复古多功能仿皮 15.6 英寸笔记本电脑包公文包斜挎包 men vintage multifunction faux leather 15.6 inch laptop bag briefcase crossbody bag。

箱包不同部位的描述关键词

扣环 buckle：一种用于连接两条带子或者皮带的装置。

拉链 zipper：用于开合袋子的口，也可用于调节口袋的大小和形状。

外部口袋 exterior pockets：指包外侧的置物小口袋。

带子 strap：用于搭载和固定背包、手提包，通常由皮革、尼龙或者帆布制成。

d 形环 d-ring：一种带有 d 字形截面的金属环，通常用于固定带子、吊绳或者钩子等配件。

扣子 clasp：一种用于开合包袋或者固定带子的装置，通常由一个纽扣和一个固定环组成。

磁扣 magnetic snap：一种使用磁力吸附的开合装置。

带扣调节器 strap adjuster：用于调节背带长度。

铆钉 rivet：用于加固包袋的各个部位。

钥匙扣 key fob：用于固定钥匙或其他小物品。

包袋脚钉 bag feet：用于保护包袋底部并防止磨损。

旋转夹 swivel clip：用于固定背带或钥匙扣等配件。

转扣 turn lock：用于固定包袋的开口。

金属铆边 metal grommet：用于加强和保护材料边缘的金属小圆环。

箱包设计 17 例

女士双肩背包

backpack design, product image, white background, women high-end multifunction soft pu leather handbag double layer large capacity backpack,floral embossed, glod grommet,blue --ar 2:3 --s 800 --v 5

women artificial leather elegant large capacity tote handbag with picasso-style decorative motifs, 3 exterior pockets --ar 2:3 --s 800 --v 5 --v 5

square backpack with rounded corners design, product image, white background, silver backpack with gold nails, crocodile skin texture, women multi-functional mini backpack with 3 exterior pockets --ar 2:3 --s 800 --v 5

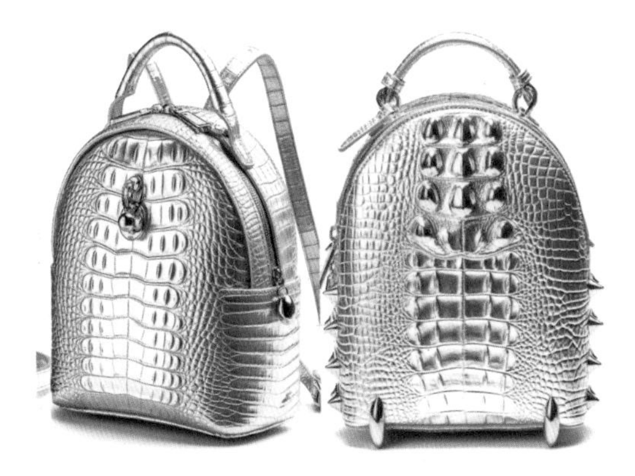

backpack design, product image,white background,silver crocodile skin texture backpack with gold nails, women multi-functional mini backpack, --ar 2:3 --s 800 --no exterior pockets --v 5

backpack design, a backpack with a face of a black cat on front, pink and red,women high-end multifunction soft pu leather handbag double layer large capacity backpack with shimmering pearls, --ar 2:3 --s 800 --v 5

双肩旅行包、帆布桶包及前卫概念包

backpack design, product image,white background,a true details realistic women's nylon water resistant travel backpack bag, red and yellow,product shot, --ar 2:3 --s 500 --v 5.2

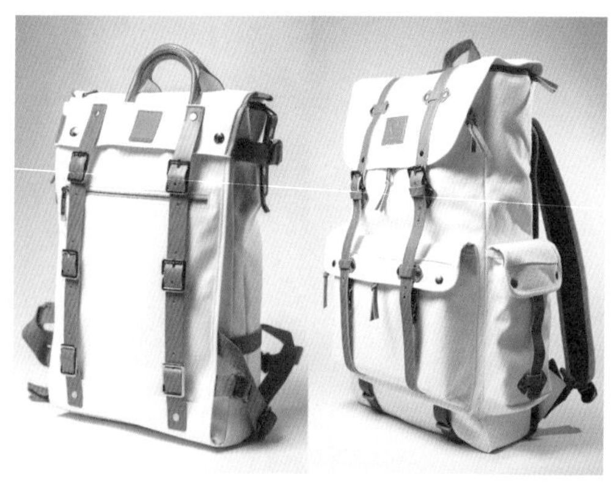

backpack design, product image，white background，square travel backpack made of polyester fabric，leather zipper pulls，backpack-style straps，orange,military style.　--ar 2:3 --s 800 --v 5　--v 5

backpack design, product image，white background，women yellow casual canvas bucket　handbag　with metal grommet，vintage style　　--ar 2:3 --s 800　--v 5

backpack design, product image，white background，women casual canvas bucket　large capcity handbag　with chinese blue print style,　--ar 2:3 --s 800　--v 5

backpack design, product image,white background,silver backpack with gold nails, crocodile skin texture , women multi-functional mini backpack with　exterior pockets　--ar 2:3 --s 800　--v 5　--v 5

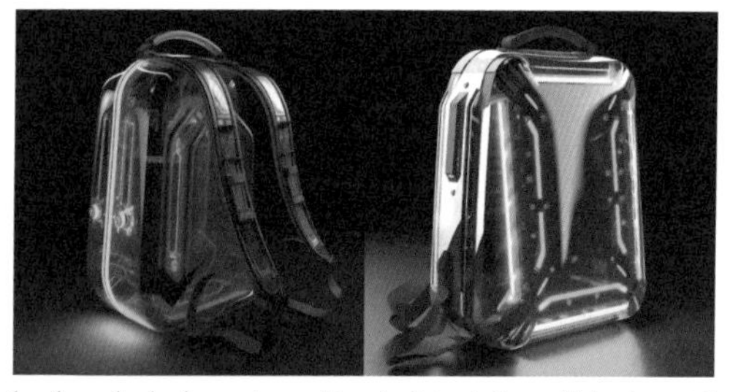

backpack design, streamlined, futuristic,polished metal alloy, transparent panel，led lighting, silver, metallic gray, electric blue, fluorescent green --s 800 --v 5.0

软壳及硬壳拉杆旅行箱

hardside luggage, silver gray , with a raised trapezoid on the front, metal pull rods, spinner wheels, made of aluminum alloy --ar 2:3 --v 5.2 --s 750

the suitcase, minimalist style,black as the main color, is embellished with red flame patterns on both sides, with diamond-shaped grooves on the surface, decorated with a few golden rivets, and has obvious leather texture on the surface,hard shell luggage with spinner wheels, aluminum alloy handle --ar 2:3 --v 5.2 --s 750

softside expandable luggage with spinners, plum, layered design, copper zipper, multiple exterior pockets --ar 2:3 --v 5.2 --s 750

luggage sets 4-piece,pink color,aerospace-grade aluminum shell,cute bear pattern, metal pull rods, spinner wheels, --ar 2:3 --v 5.1 --s 750

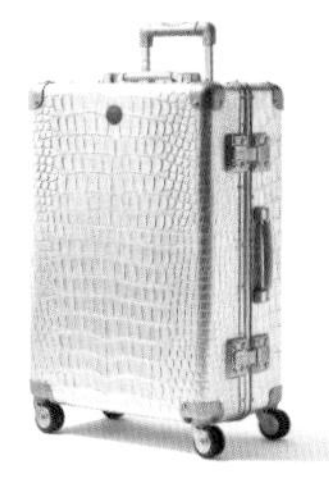

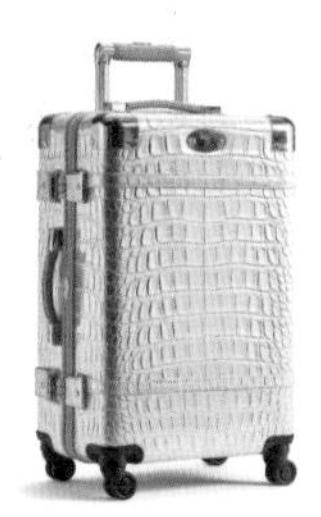

luggage design, white background,silver hardside with gold nails, crocodile skin texture,multi-functional with exterior pockets,metal pull rods, spinner wheels --s 800 --v 5.0

luggage design, white background,irregular polyhedron structure shape,made of glossy composite, holographic projection,biometric handle,laser blue and silver --s 800 --v 5.0

第 7 章 Midjourney 服装鞋子设计 29 例

常见服装及鞋设计关键字

不同服装关键词

连帽衫 hoodie、棒球夹克 baseball jacket、羽绒服 down jacket、风衣 trench coat、针织衫 sweater、运动夹克 track jacket、牛仔夹克 denim jacket、衬衫 shirt、西服 suit、短袖 t 恤 short-sleeved t-shirt、长袖 t 恤 long-sleeved t-shirt、圆领 t 恤 crew neck t-shirt 、马甲 vest、短裤 shorts、运动裤 athletic pants、雨衣 raincoat、皮衣 leather jacket、羽绒服 down jacket、牛仔衣 denim jacket、牛仔裤 denim jeans、毛衣 sweater、针织衫 knitted sweater、晚礼服 evening dress、中山装 zhongshan suit、唐装 tang suit、工作服 work uniform、迷彩服 camouflage uniform、汉服 hanfu、polo 衫 polo shirts、打底裤 leggings、旗袍 cheongsam。

不同服装部位关键词

领口 collar、领带 necktie、翻领 lapel、扣子 button、袖口 cuff、袖子 sleeve、腰带 belt、裤腰 waistband、裤腿 trouser leg、衬衫领 shirt collar、衬衫袖 shirt sleeve、胸部 chest、胸口袋 chest pocket、拉链 zipper、扣子 button。

不同鞋子关键词

运动鞋 sports shoes、高跟鞋 high heels、皮鞋 leather shoes、凉鞋 sandals、靴子 boots、休闲鞋 casual shoes、帆布鞋 canvas shoes、拖鞋 slippers、船鞋 boat shoes、雨鞋 rain boots、军鞋 army boots、洞洞鞋 hole shoes、椰子运动鞋 yeezy foam shoes、牛津商务休闲鞋 oxfords classic casual dress shoes、马丁靴 dr. martens shoes、发光鞋 light-up shoes、细高跟鞋 stiletto heels、粗跟鞋 chunky heels、坡跟鞋 wedge heels、露趾高跟鞋 peep-toe heels、短靴高跟鞋 short boots with heels、芭蕾舞鞋 ballet heels。

不同鞋子部位关键词

鞋面 upper、鞋头 toe、鞋舌 tongue、鞋垫 insole、鞋底 outsole/sole、中底 midsole、鞋眼孔 eyelets、鞋带 lace、鞋跟 heel、鞋盒 box。

不同鞋子结构设计关键词

运动鞋 sneakers/sports shoes，透气网状鞋面 breathable mesh upper、缓震鞋底 cushioned sole、弧形设计的鞋跟 arched heel。

高跟鞋 high heels，尖头鞋头 pointed toe、高跟鞋跟 stiletto heel、鞋面材质 various materials。

靴子 boots，高筒靴 high boots/over-the-knee boots、短靴 ankle boots、防水材质 waterproof materials、拉链或扣子 fastening with zipper/buckle。

滑板鞋 skate shoes，圆头鞋头 round toe、防滑鞋底 grippy sole、舌头鞋舌 padded tongue、鞋面材质 suede or canvas。

人字拖 flip-flops/thongs，y 型鞋带 y-shaped straps、简单的鞋面设计 simple upper design、柔软的鞋底 soft sole。

乐福鞋 loafers，极简鞋头设计 simple toe design、无鞋带 no laces、平底设计 flat sole。

经典运动鞋 classic sneakers，经典的鞋头设计 classic toe design、简单的鞋身设计 simple upper design、耐磨鞋底 durable sole。

拉链靴 zipper boots，便捷的拉链设计 convenient zipper design、多样的鞋面材质 various upper materials、平跟 flat heel。

凉鞋 sandals，简单的鞋面设计 simple upper design、露出脚趾的设计 open-toe design、可调节的鞋带 adjustable straps。

服装及鞋设计 29 例

运动服及 polo 衫

new set of red and black sport uniform, inspired from chinese dragon --s 1000 --q 2 --ar 16:9 --v 5

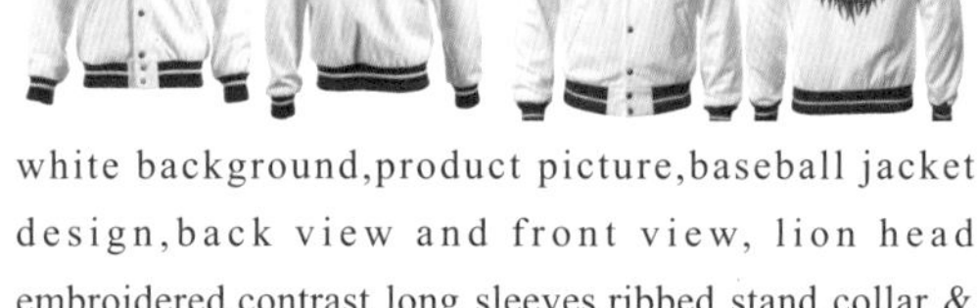

white background,product picture,baseball jacket design,back view and front view, lion head embroidered,contrast long sleeves,ribbed stand collar & cuffs,green and white --ar 16:9 --s 800 --no model --v 5

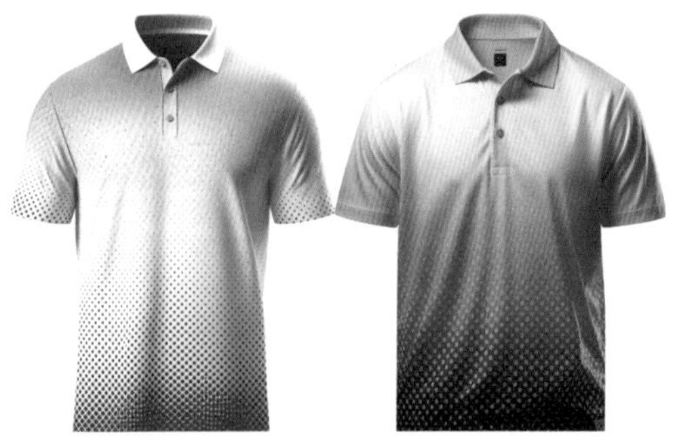

white background,product picture polo shirts design,many black gradient dots gradually fade away into red , geometric style. --ar 2:3 --s 800 --no model --v 5

polo shirt design, product view, white background , white and black stripes on the chest position --ar 2:3 --v 5.2 --no human --s 750

连帽衫与户外服装

3d mocha unisex hoodie in 3d front position, features tribal repetitive patterns in cream. white isolated background --no human,text --ar 2:3 --v 5.2

3d purple men's hoodie in 3d front position, features a repeating diamond pattern in yellow. white isolated background --no human,text --ar 2:3 --v 5.2

storm jacket, product view, loose ， a-line fit clothes, small pocket ， splicing process， rock gray, dark brown and olive contrast --ar 2:3 --v 5.2 --no human --s 750

a softshell jacket for man,several pockets ， product view, white background , loose， a-line fit，very diverse，dark brown and olive ,small line pattern on bottom edge --ar 2:3 --s 500 --v 5.2

前卫概念设计服装

fashion model wearing a subversive stylish costume , white background --ar 2:3 --s 600 --v 5

plastic waste fashion clothes , fashion photo,white background , --ar 2:3 --s 600 --v 5

针织衫与羽绒服

fashion model wearing a subversive stylish costume , white background --ar 2:3 --s 600 --v 5

down jacket,product view, white background, asymmetrical pockets,multi-layer,cut and line design, fit and streamlined, detachable hat or collar , black contrasting with charcoal gray --ar 2:3 --v 5.2 --no human

中式旗袍

chinese style ancient red wedding dress, surreal, high definition,embroidery craftsmanship,red with gold trim,abundant florals --v 5.1 --s 750

chinese ancient hanfu clothing, product view, white background ,ming dynasty ,embroidery craftsmanship,red with gold trim,abundant florals ,bright colors, embroidery, chinese dragon and phoenix --v 5.2 --s 750

短袖与长袖 T 恤

colorful hawaiian printed shirts,product view, white background, in the style of baroque nu-vintage, rusticcore, vintage aesthetics

long-sleeved t-shirt design, product view, white background , great horned owl in circle , --ar 2:3 --v 5.2 --no human --s 750

高跟鞋与女士凉鞋

the most luxurious designer style leather skinny heel stilettos,diamonds,gold embellishment,fashionable,close up shot --ar 3:2 --v 5.2 --s 750

a pair of exquisite crystal stiletto heels, shimmering with elegance and sophistication, featuring transparent straps adorned with sparkling rhinestones --ar 3:2 --v 5.2 --s 750

silver wedding dress shoes, white background

a leather gladiator woman sandals with cut out detailing and a cowboy heel, bold, with 6 leather straps but only 3 metal silver buckles , black lines, --v 5.2 --s 750

women's platform sandals wedge open toe ankle strap lace wedding shoes bridesmaid shoes,ribbon, lace, decorated with diamonds, --ar 3:2 --v 5.2 --s 750

women's low heel flat lolita shoes,t-strap round toe ankle strap,with cute bear engraved texture pattern, yellow and black leather weave --ar 3:2 --v 5.2 --s 750

运动鞋

sports shoes design ,product picture,white background,piet mondrian style,complex and exquisite shoe upper structure design. --ar 3:2 --s 500 --v 5

shoes design ,product picture,white background,the upper of the shoe is designed with a lightweight and breathable mesh, while the sole is made of durable rubber material. it features a staggered hole-like cushioning structure, in blue and white color. --ar 3:2 --s 500 --v 5

sports shoes design,white background, product image, breathable lace pattern on the upper, and a white translucent plastic sole with shiny led lights, pink color. streamlined design. --ar 3:2 --s 600 --v 5

帆布鞋

white background,product picture ,vans shoes design,low platform slip on shoes,lion, pop art retro comic style --ar 3:2 --s 800 --v 5

white background,product picture ,vans shoes design,low platform slip on shoes,black white checkerboard --ar 3:2 --s 800 --v 5

前卫概念设计感鞋

white background, product image,sports shoe design, miniature art style, dragon-shaped upper, with a thick sole carved with a complete dragon pattern, exuding a strong heroism. elegant decorative design, highly detailed, --ar 3:2 --s 600 --v 5

futuristic luxury shoe designed by louis vuitton --q 2

第 8 章　Midjourney 不同风格插画绘制 60 例

两种方法生成插画图像

提示语法

在提示语中添加 2D 二维平面图像、插画 illustration、 线描 line art、 手绘 hand drawn、矢量图 vector、绘画 drawing、水彩画 watercolor、铅笔画 pencil 、水墨风格 ink style、动画 anime、平面绘画 flat painting 等明确指出图像类型属于插画、绘画类的词语，或者在提示语中使用 in style of 、by ……语句，并在后面添加了插画艺术家的名字，则可以轻松得到各种不同类型的插画图像。

例如生成下面的图像时，笔者添加了细线条 thin lines、矢量图像 vector image、抽象线条 abstract lines 三个关键词，并使用了极简风格肖像 minimalist portrait，因此，生成的整体风格与效果与构想中的图像基本没有太大出入。

a minimalist portrait of a woman wearing a hat and scarf with tapered lines on a dark red gradient background,simple, thin lines, vector image, abstract lines --ar 2:3 --v 4

生成下面的插画时，笔者在提示语中加入了知名插画艺术家的名字 peter elson，他是英国知名科幻插画家，其作品的主题常常围绕着复杂的机器、外星人、星球和宇宙船等素材。

ci-fi worldly garden of paradise by peter elson --ar 2:3 --s 800 --v 5

参数法

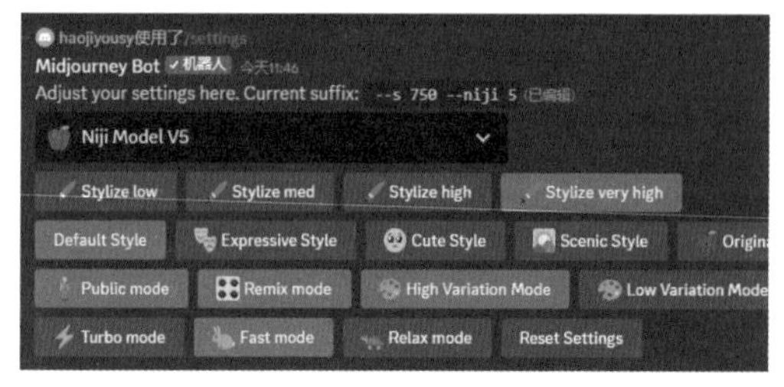

在第 1 章曾经讲解过的 niji 模型，专门用于生成插画。

撰写提示语时，还可以添加使用 --style cute、--style expressive、--style original 以及 --style scenic 参数，以控制生成的效果。

pokemon gym leader fan character concept,full portrait, fairy type pokemon, inspired by xernieas and sylveon, cute, light skin, heterochromia, long thick white pinkish colored hair, pink and white colorful and vibrant, auroracore, by studio trigger --ar 2:3 --niji 5 --s 750

理解版本参数与插画的影响

虽然无论使用 v4、v 5、v5.1、v5.2 还是 niji 版本参数，均可以生成插画，但正如本书在讲解版本参数时所指出的，v5.2 更偏照片、写实，因此，在使用时不能因为 v5.2 版本级别更高，就偏向于使用此版本参数。

下面展示的是，在使用同一组提示语的情况下，分别使用不同的版本的效果图。

--v 5.2

--v 5.1

--v 5

--niji 5

不同插画风格 60 例

中式剪纸平面风格效果

chinese new year posters ,red ,sunset, in the style of minimalist stage designs, landscape-focused,heavy texture --ar 2:3 --v 5.2 --s 350

细节华丽风格效果

eguchi hisashi style,a captivating portrait of a peking opera actress, expressive eyes and dramatic features, colorful costume --niji 5 --s 800 --ar 2:3

迷幻风格效果

psychedelic, sci-fi, colorful, disney princess --ar 2:3 --v 5

华丽植物花卉风格效果

fusion between pointillism and alcohol ink painting, vibrant, glowing, ethereal elegant goddess by anna dittmann, baroque style ornate decoration, curly flowers and branches，metallic ink, --ar 2:3 --v 4

滑稽风格插画效果

caricature art in the style of david low --v 5.1 --s 500

水彩画风格效果

aerial view, watercolor,a pirate stands on a very high hill, looking down at the whole city,in style of anders zorn --ar 9:16 --v 4

素描效果

aerial view, graphite sketch ,many detail,a pirate stands on a very high hill, looking down at the whole city --ar 9:16 --v 4

三角形块面风格效果

aerial view,in the style of cubist multifaceted angles, dark green and blue, many detail,a pirate stands on a very high hill, looking down at the whole city --ar 9:16 --v 4

彩色玻璃风格效果

aerial view, vibrant stained glass, in style of john william waterhouse,many detail,a pirate stands on a very high hill, looking down at the whole city --ar 9:16 --v 4

cubist multifaceted angles：这是一种与立体派 cubism 相关的艺术风格。这种风格是将一个物体分解成多个几何形状和角度，然后对他们重新组合，以同时呈现多个角度。从视觉效果上看，使用这种风格绘制的图像是一个复杂、多面的形象，具有极强的块面化特点。

stained glass：即彩绘玻璃风格，它模仿了彩绘玻璃窗的效果。这种风格使用明亮、饱和的色彩和大胆的线条来营造一种富有活力和生命力的感觉。这种风格通常与教堂的彩绘玻璃窗联系在一起。在当代艺术中，它也被广泛应用于其他领域，如插画、海报、装饰艺术等。

剪影画效果

aerial view, line draw, in style of silhouette,many detail,a pirate stands on a very high hill, looking down at the whole city --ar 9:16 --v 4

黑白线条画效果

aerial view, coloring book page, simplified lineart vector outline ,many intricate details,illustrator,a pirate stands on a very high hill, looking down at the whole city --ar 9:16 --v 4

扎染风格效果

aerial view, tie dye illustration ,many detail,a pirate stands on a very high hill, looking down at the whole city --ar 9:16 --v 4

立体主体拼贴效果

aerial view, a pirate stands on a very high hill, looking down at the whole city,cubist screen print illustration style, by albert gleizes and juan gris style --ar 9:16 --v 4 --s 400

tie dye：即扎染，是一种传统的染色工艺，也称为“绑染”或“结染”，其技术是将织物折叠、卷起、绑扎成各种形状，再将其浸入染料中，从而形成具有独特图案和纹理的织物。

扎染的起源可以追溯到公元前 2000 年左右的古代印度，后来在中国和日本也逐渐形成了独特的传统染色工艺，被广泛应用于传统和现代服装、家居用品、工艺品和艺术品等方面，其独特的图案和纹理深受大众的赞赏和喜爱。

cubism：即立体主义，是一种现代艺术风格，起源于 20 世纪初期的法国，由毕加索和布拉克等人发明并发扬光大。立体主义的主要特征是以几何体为基础，将被描绘的物体分解为几何形体，再将其用立体角度组合，强调空间感和视觉效果。这种风格主要运用于绘画和雕塑中，对当时的现代艺术潮流产生了深远影响。阿尔贝·格莱兹 albert gleizes 和胡安·格里斯 juan gris 是法国立体派运动的重要代表人物。

黑白色调画效果

aerial view, concept art ,black line art work,coloring page for adult,black and white, many detail,a pirate stands on a very high hill, looking down at the whole city --ar 9:16

霓虹风格效果

aerial view, light painting neon glowing style, many detail,a pirate stands on a very high hill, looking down at the whole city --ar 9:16 --v 4

战锤游戏风格效果

aerial view, warhammer style, dramatic lighting , many detail,a pirate stands on a very high hill, looking down at the whole city --ar 9:16 --v 4

波普艺术复古漫画风格效果

aerial view, in pop art retro comic style, in style of roy lichtenstein,illustration, many detail,a pirate stands on a very high hill, looking down at whole city --ar 9:16 --v 4

晕染插画效果

aerial view, a pirate stands on a very high hill, looking down at the whole city, watercolor painting, impressionist style, rich and muted tones, diffused lighting, atmospheric and expressive, wet-on-wet technique --s 550 --v 5.0

霓虹平面几何插画效果

aerial view, a pirate stands on a very high hill, looking down at the whole city,by kazumasa nagai, electric neon colors in a dazzling mood ,in the style of camille walala --s 750 --v 5.1

粉笔画效果

aerial view,chalk drawing,white lines on black background,many detail,a pirate stands on a very high hill, looking down at the whole city --ar 9:16 --v 4

炭笔画效果

aerial view,charcoal drawing,black and white,many detail,a pirate stands on a very high hill, looking down at the whole city --ar 9:16 --v 4

点彩画效果

aerial view,pointillism,in style of georges seurat,many detail,a pirate stands on a very high hill, looking down at the whole city --ar 9:16 --v 4

构成主义风格效果

aerial view,constructivism,in style of piet mondrian,many detail,a pirate stands on a very high hill, looking down at the whole city --ar 9:16 --v 4

极简平面插画效果

aerial view, a pirate stands on a very high hill, looking down at the whole city,line art 2d illustration style, organic shapes, minimalist --s 750 --v 4

模拟儿童绘画插画效果

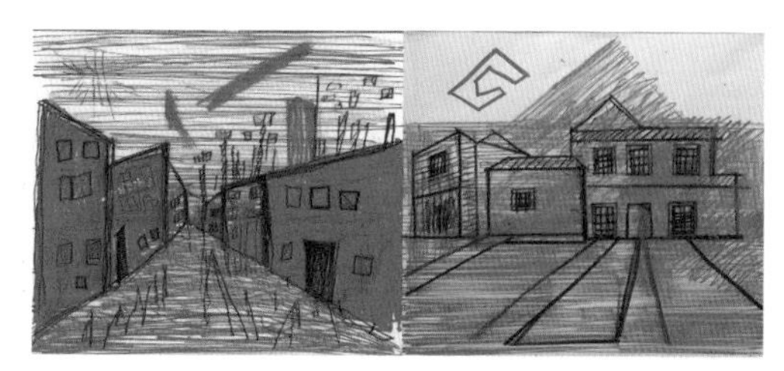

a low quality child's red ink ball-point pen drawing of school in the style of primitive linear drawing by a 4 year old kid, simple irregular lines, mistakes --v 5.2

反白轮廓插画效果

white silouette,illustrator, middle ages elf warrior, full portrait,pure black background --ar 9:16 --v 4

像素化效果

aerial view,8bit,a pirate stands on a very high hill, looking down at the whole city --ar 9:16 --v 4

木刻版画效果

aerial view,mabel annesley style, woodcut print ,many detail,a pirate stands on a very high hill, looking down at the whole city --ar 9:16 --v 4

油画效果

aerial view, a pirate stands on a very high hill, looking down at the whole city,oil painting, brush strokes, by razumovskaya, --ar 9:16 --v 4 --s 800

装饰彩条拼贴插画效果

dancer,abstract, op art fashion, multiple layers of meaning, paisley and ikat, origamic tessellation, vibrant brush strokes, intricate patterns, thought-provoking visuals --s 500 --v 5.1

厚涂笔触风格插画效果

fusion between sgraffito and thick impasto, stunning surreal sun flower art --s 500 --v 5.1

细腻植物插画效果

peach tree branch, botanical illustration, white background, style of margaret mee --ar 16:9 --q 4

超现实主义水彩风格插画师

colored loki with white horse surrounded by blue flame stencil drawing for tattoo,in style of leonardo da vinci, --ar 2:3 --v 5 --q 2 --s 750

超现实立体主义风格插画师

abstract,in style of pablo picasso, oilpainting, musikinstrument, thick paint, expressive --v 4 --ar 2:3 --q 5 --s 500

一笔画极简风格插画

one line art style, woman face illustration with flowers --s 750

滴溅效果风格插画

chinese dragon shaped rainbow splash vector art illustration --s 750

抽象插画效果

abstract pattern in black on white background with shapes, lines and dots, in the style of jack hughes, free brushwork, rounded forms, expressive characters, mike winkelmann, illustration, goro fujita --ar 2:3 --v 5.2

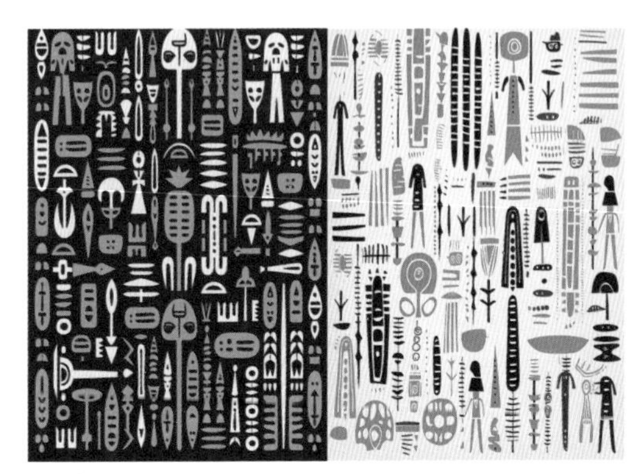

simple clear distinct styled seamless pattern with tribal elements, in the style of subtle, earth tones, igbo (ibo) art, quirky shapes, mustard tones, hand-drawn elements--tile --s 250 --v 5.2 --c 10

abstract circles,rectangles, overlapping, intersecting, black and white, bauhaus, minimalist, poster --ar 2:3 --s 750

line art 2d illustration in the style of google illustration, wavy flowing lines, organic shapes, spheres, dots, splatter, minimalist --ar 2:3 --s 750

涂鸦艺术风格插画

dancer by keith haring. flat graphic ,white background. --s 750 --v 5.1

凡·高笔触效果风格插画

in the style of van gogh's starry sky, a treehouse in valley，--s 750 --v 5.1 --s 750

“找找看”风格插画效果

generate a 'where's waldo?' style illustration. the scene should be set in village with an abundance of characters and objects.. --v 5 --ar 3:2 --s 750

20 世纪 50 年代插画效果

vintage 50's advertising illustration of futuristic trip journey on a surreal alien planet, alien ruins jungle, robots, gorgeous women, --ar 9:16 --v 5.2

vintage 50's advertising illustration of futuristic music studio interior, surreal instruments with keyboards, surreal beat machines, robots, gorgeous women, --ar 9:16 --v 5.2

水墨画及工笔画效果

chinese ink painting, wu guanzhong style, under a willow tree, two boys of about three years old, dressed in chinese han clothes, squatting on the ground to play --ar 2:3 --v 5.2

two people sitting on a rock,pen and ink --ar 2:3 --v 5.2 --s 750

under the huge pine tree, dry bursh style,ancient chinese old man and the child . fog background, a lot of white space, traditional chinese ink painting style. --ar 2:3 --s 450 --style raw --v 5.2

birds，flowers，by emperor huizong of song dynasty --ar 9:16 --v 5

acrylic art, art by yoji shinkawa, kungfu character --ar 2:3 --s 750 --v 5.2

acrylic art, art by yoji shinkawa, kungfu character --ar 2:3 --s 750 --niji 5

四格漫画效果插画

superman fight spideman, fight, comic pages --ar 1:2 --v 4

插画教学示范效果图像

step-by-step guide for draw,teach book page layout,the female fighter character design,fight pose,full portrait, white background, --q 4

涂色书白描效果插画

coloring book page, simplified lineart vector outline,chrysanthemums,many intricate details ,clean, white background --ar 2:3 --s 850 --q 2 --v 5

示意草图效果插画

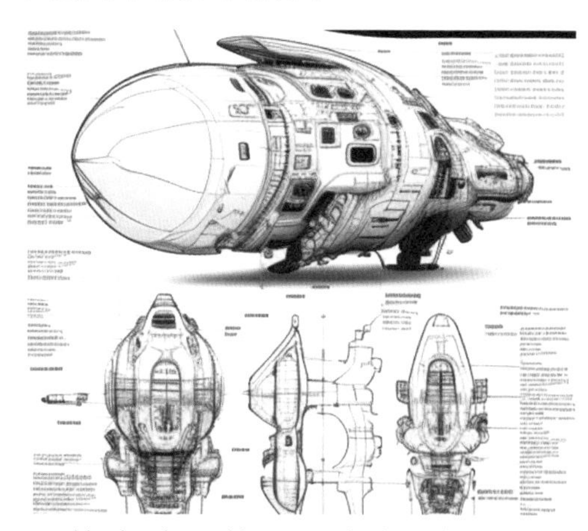

spaceship drawing guide, vector, intricate details --s 500 --v 4

同一角色多角度效果插画

long hair female warrior close up character design, concept design sheet, white background

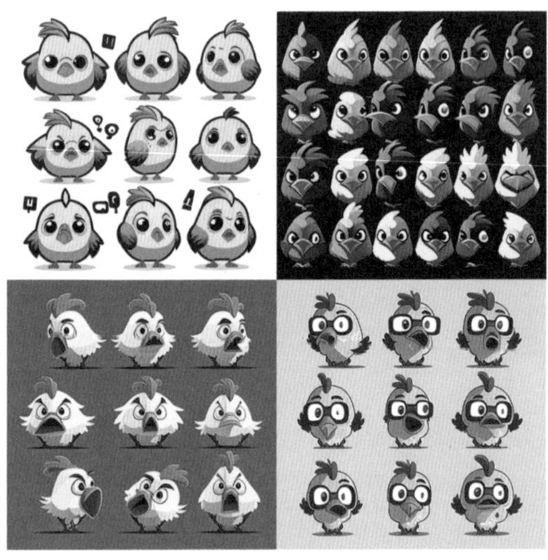

cute chicken vector character facial expression sheet

adorable toddler siamese kitten cartoon, dynamic poses, character pose sheet, watercolor, white background --v 5

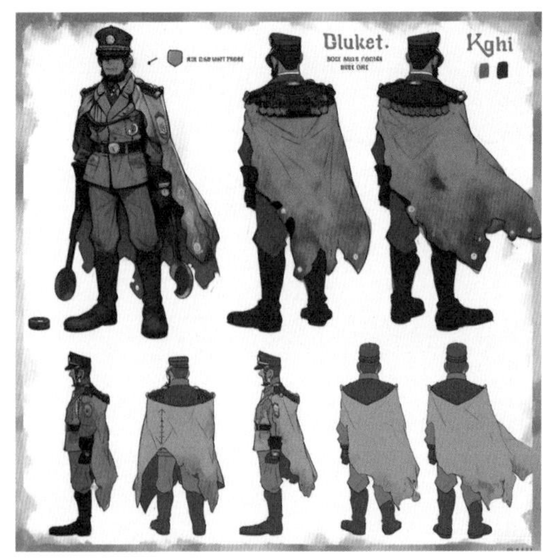

old dieselpunk king ,turnaround sheet,detailed, watercolor, white plain background --niji 5

第 9 章 Midjourney 珠宝设计 75 例

珠宝设计常用关键词

常见的珠宝类型提示关键词

戒指 ring、手链 bracelet、项链 necklace、耳环 earrings、颈链 choker、腰链 waist chain、脚链 anklet、戒指套装 ring set、项链套装 necklace set、个性化首饰 personalized jewelry、珠宝耳钉 stud earrings、耳坠 drop earrings、手镯 bangle、护身符珠串 beaded necklace、耳环 earrings、手镯 bracelet、把件 hairpin、佩饰 pendant、钗子 hairpin with tassel、坠子 pendant with tassel、玉佩 jade pendant、戒指 ring、镯子 bangle、手链 bracelet、首饰套装 jewelry set、指环 finger ring、璎络 tassel pendant、花翎 hair ornament with flowers、蝴蝶结 bowknot ribbon、卡子 hair clip、耳坠 earring drop、头环 headband、腰坠 waist pendant、腕坠 wrist pendant、头饰 headwear。

常见的珠宝材质提示关键词

黄金 gold、白金 platinum、银 silver、钻石 diamond、珍珠 pearl、翡翠 jade、红宝石 ruby、蓝宝石 sapphire、绿宝石 emerald、玛瑙 agate、水晶 crystal、琥珀 amber、玛雅石 lapis lazuli、红玛瑙 carnelian、绿松石 turquoise、黑珍珠 black pearl、珊瑚 coral、玻璃 glass、玫瑰金 rose gold、白银 sterling silver、黑陶瓷 black ceramic、老坑翡翠 old mine jadeite、蛋白石 moonstone、石榴石 garnet。

知名珠宝品牌提示关键词

卡地亚 cartier、蒂芙尼 tiffany、宝格丽 bvlgari、汉利·温斯顿 harry winston、梵克雅宝 van cleef & arpels、万宝龙 montblanc、伯爵 piaget、萧邦 chopard、施华洛世奇 swarovski、爱马仕 hermès、卡尔文·克莱恩 calvin klein、戴比尔斯 de beers、范思哲 versace。

地域风格关键词

中国风 chinese style、西藏风 tibetan style、日本风 japanese style 、印度风 indian style 、伊斯兰艺术 islamic art、波斯艺术 persian art、古埃及艺术 ancient egyptian art、古希腊风 ancient greek style 、古罗马风 ancient roman style 、巴洛克风格 - baroque style 、古埃及风 ancient egyptian style、非洲部落艺术 african tribal art、非洲现代艺术 african modern art 、印第安艺术 native american art、美洲艺术 native american modern art、古代印加文化艺术 ancient inca cultural art、阿兹特克艺术 aztec art、玛雅艺术风格 mayan art style、墨西哥民间艺术 mexican folk art、印加艺术风格 inca art style。

珠宝设计 75 例

不同文化地域风格设计

earring jewelry design, indian style, delicate, elegant, detailed intricate, photorealistic, product view, --s 150 --v 5

earring jewelry design,tibetan style,delicate, elegant, detailed intricate,photorealistic,product view, --s 150 --v 5

earring jewelry design,arabian style,delicate,elegant, detailed intricate,photorealistic, product view, --s 150 --v 5

gold and diamond necklace design,aztec art style,perfect symmetry minimalist style, simplified lines and shapes --v 5.2 --s 450

necklace design,gold and diamond,mexican folk style,perfect symmetry minimalist --v 5.2 --s 650

necklace design,gold and diamond ,mayan style,perfect symmetry minimalist --v 5.2 --s 850

earring jewelry design,chinese style,delicate, elegant, detailed intricate,photorealistic,product view, --s 150 --v 5

不同设计师风格设计

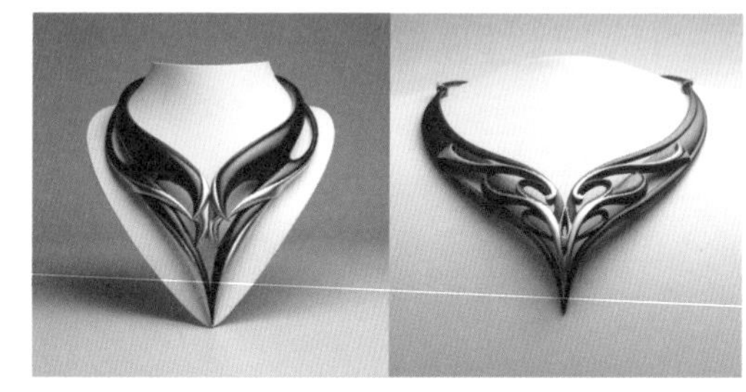

in style of zaha hadid,necklace jewerly design,product view,minimalist --v 5.2 --s 550

in style of frank gehry, necklace jewerly design, product view --v 5.2 --s 750

in style of paul brandt, jewelry design, necklace design, gemstones and diamonds, luxury, delicate, elegant, detailed intricate, photorealistic, product view, --s 150 --v 5

in style of wolfers frères, jewelry design, necklace design, gemstones and diamonds,luxury, delicate, elegant, detailed intricate,photorealistic, product view, --s 150 --v 5

in style of pierre sterlé, jewelry design, necklace design, gemstones and diamonds, photorealistic, product view, --s 150 --v 5

in style of donald claflin, jewelry design, necklace design, gemstones and diamonds, photorealistic, product view, --s 150 --v 5

in style of lacloche frères, necklace jewerly design, product view --v 5.2 --s 750

不同拟物风格设计

delicate pagoda shape pendant jewelry designs, silver and gold, minimalist --v 5.2 --s 550 --style raw

jewelry design, earring design, earrings, classical, rococo period, silver, details, summer, eagle,wings, lattice

necklace design, jewelry design, native american style, incorporate feathers, indian totems and coral elements, gemstone centerpiece, vibrant stone, delicate and sturdy gold chain, sparkling gemstone accents, simple yet eye-catching design --s 150 --v 5

earring jewelry design, indian style, the jewelry shape is inspired by the lotus shape element delicate, elegant, detailed intricate,photorealistic, product view, --s 150 --v 5

earring jewelry design, chinese style, the jewelry shape is inspired by the minimalist style of dragon and phoenix elements, delicate, elegant, detailed intricate, photorealistic, product view, --s 150 --v 5

ancient chinese jade ruyi shaped jewelry pendant design, silver and gold, minimalist --v 5.2 --s 550 --style raw

jewelry design sakura ring, gold, realistic delicate design, soft illumination, dreamy, fashion

necklace design, jewelry design, diadem with phoenix bird, incorporate feathers elements, ruby and obsidian, detailed intricate, photorealistic, product view, --s 150 --v 5

jewelry design, ring design, pearls and diamonds, small heart shape, delicate, elegant, detailed intricate, photorealistic, product view, --s 150

earring jewelry design, sapphire and red gem, stars and moon, delicate, elegant, detailed intricate,photorealistic, product view, --s 150

jewelry design, a close up of a brooch, shape is dance woman in a dress, classic dancer striking a pose, diamonds, photorealistic, product view, --s 550 --v 5

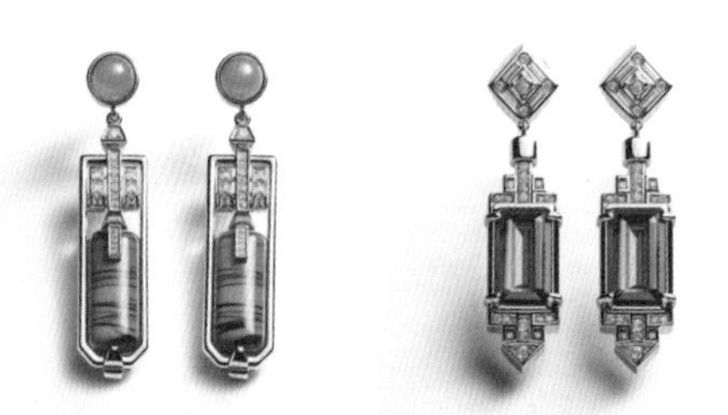

earrings, chinese window, gold, rectangular, vivid green chalcedony, diamonds luxurious minimalist style --v 5.2 --s 350

earring design, hollow gourd shape, perfect symmetry minimalist style, simplified lines and shapes --v 5.2 --s 350 --style raw

a silver dove necklace, in the style of minimalist impressions, white and aquamarine --v 5.2 --s 550 --style raw

simple strokes chinese dragon style pendants, sapphire main stone, small pearl, diamond, gold, platinum, minimalist style, vertical length --v 5.2 --s 50 --style raw

不同知名 IP 概念设计

a jewelry design, alien themed ring, gemstones and diamonds, luxury, delicate, elegant, detailed intricate, photorealistic, product view, --s 150 --v 5

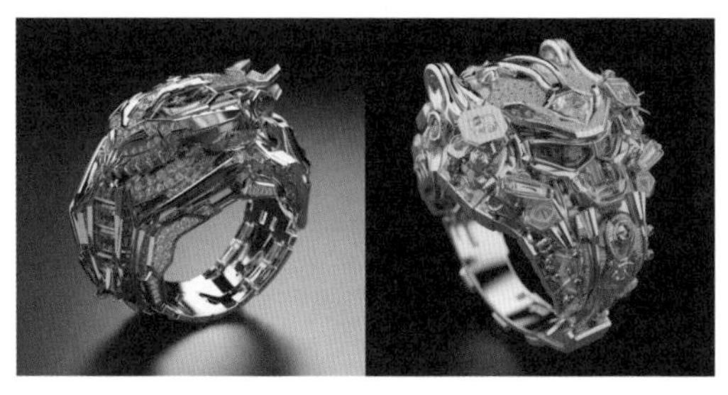

jewelry design,transformers themed ring, gemstones and diamonds, luxury, delicate, elegant, detailed intricate, photorealistic, product view, --s 150 --v 5

jewelry design, barbie themed ring, gemstones and diamonds, luxury, delicate, elegant, detailed intricate, photorealistic, product view, --s 150 --v 5

school badge style, harry potter style ring, with diamonds and gold, high jewelry, white background --v 5.2 --s 750

architectural design, the starry night van gogh pendant, with diamonds and gems, high jewelry, --v 5.2 --s 750

pendant with leonardo da vinci's mona lisa, tiny ring with diamonds and gems, high details, white background --v 5.2 --s 750

jewelry pendant, chinese kung fu shaolin temple concept design, jade and gold, minimalist --v 5.2 --s 550 --style raw

不同几何体造型设计

teardrop ripples ring, the ring's design mimics the ripple effect caused by a teardrop falling into water, with each ripple adorned by tiny diamonds, gold and diamond, white background --v 5.2 --s 750

spiral dance earrings, the earrings elegantly blend spiral lines in their shape, gold and diamond, white background --v 5.2 --s 750

earring made of gold, the circle with a diameter of three centimeters is shiny on the circle, and there are raised grooves on the circle, --v 5.2 --s 750

white background, flat disc shape jewelry pendant with diamond chain, the interior is a thick line of rotation, hollow design, rubies are inlaid in each of the four corners of the square, and diamonds are inlaid in other places

white background, rectangle jewelry pendant, the interior is a thick swirl line made of diamond, hollow design, rubies are inlaid in each of the four corners of the square, and diamonds are inlaid in other places --v 5.2 --s 750

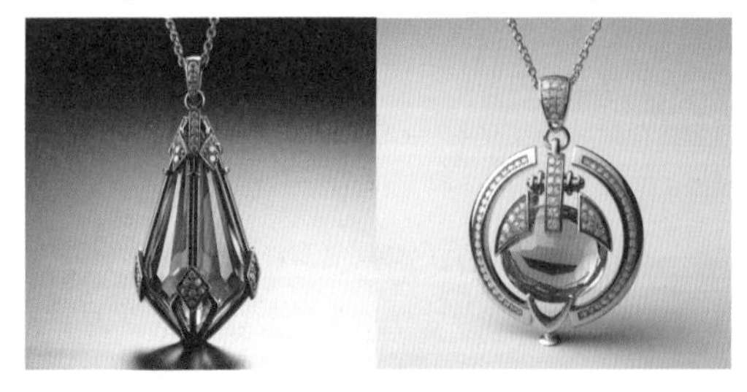

a pink crystal pendant with a gold frame on it, geometric constructivism, in the style of layered lines woven perforated, --v 5.2 --s 750

multi-triangle necklace, the necklace is composed of a series of triangles of varying sizes, arranged in multiple layers, creating a gradient flow. --v 5.2 --s 750

dangling droplet earrings, the shape of these earrings resembles the posture of a droplet falling from a height, with slender and smooth lines, white background --v 5.2 --s 750

pendants,the milky way style, sapphiremain stone,little sapphire, little pearl, diamonds, gold and white gold --v 5.2 --s 750

infinity sign pendant diamond jewels, minimalist, elegant --v 5.2 --s 750

earring made of gold, the circle with a diameter of three centimeters is shiny on the circle, and there are raised grooves on the circle, --v 5.2 --s 750

pyramid stacking necklace,the necklace's design resembles a stack of pyramids, gradually increasing in size from top to bottom,minimalist --v 5.2 --s 750

circular square necklace,a series of circular square shapes hang on the chain, each square connected at a different angle, forming a continuous circular structure, white gold and diamond --v 5.2 --s 750

hexagram pattern necklace，the center of the necklace is formed by six equilateral triangles, creating a perfect hexagram pattern. each angle of the hexagram is adorned with a small circular inset，gold and diamond --v 5.2 --s 750

trendy luxury diamond earrings, trompe l'oeil pleated style, inspired by the tail fins of betta fish, minimalist style, linear design --v 5.2 --s 350 --style raw

gold and diamond earring design, coral shape, minimalist style, simplified lines and shapes, abstract generalized style --v 5.2 --s 350 --style raw

a pair of emerald jewels, in the style of organic shapes and curved lines, twisted branches, nature-inspired pieces, minimalist,fashion --v 5.2 --s 350 --style raw

pendant jewelry design, a ring is divided into 8 equal parts, each part is inlaid with a ruby, the middle of the ring is a big diamond, flat design minimalist,fashion --v 5.2 --s 350 --style raw

a gold ring with an oval cut diamond in a swirl design, in the style of melting pots, minimalist --v 5.2 --s 750

necklace in cartier style, full moon and star, minimalism, fashion --v 5.2 --s 350 --style raw

horizontal bar pendant with diamond at the bottom, gold,minimalist art deco --v 5.2 --s 750

rectangle and sphere pendant ， tiny,elegant ,with different shape and color gems white background, --v 5.2 --s 750

regal radiance earrings ruby gold hoops pearl dangles crystal chandeliers diamond drops silver huggies silk tassels satin jacket ruby cluster gold ear cuffs

a stunning shot of kohinoor chandelier earrings large chandelier rose gold, diamond textured asymmetrical vintage and luxurious diamond against the metal kohinoor diamond accents,

haute joillerie mens pendant cross made from white gold on leather strap, cylinder base shape of the cross legs, in the center a blue diamond, encrusted with small white diamonds, --v 5.2 --s 750

pendant cross , in the center a big blue diamond, encrusted with small white diamonds, maskulin reduced design with modern gothic touch, --v 5.2 --s 750

opulent whispers earrings tanzanite gold hoops pearl dangles crystal chandeliers citrine drops gold huggies silk tassels silk jacket amethyst cluster silver ear cuffs

grandiose magnificence earrings sapphire rose gold hoops pearl dangles crystal chandeliers diamond drops gold huggies silk tassels satin jacket sapphire cluster silver ear cuffs

a stunning shot of an earring brass with coral and shell inlay ,stud,coral --v 5.2

a stunning shot of an earring ,textured and colorful , bohemian beadwork --v 5.2

saturn pendant jewelry, marguerite style, three-dimensional surround structure, delicate structure, delicate and slender elements --v 5.2 --s 750

pendant with mandala and swilr line , ring with diamonds and gems， high details，white background --v 5.2 --s 750

stacking gold swirl shaped ring with rubies emeralds diamonds --v 5.2 --s 750

planet idea ,sunrise and moon impression pendant,with diamonds and gems,high jewelry,white background , --v 5.2 --s 750

architectural design，sunrise impression pendant，with diamonds and gems，high jewelry，white background --v 5.2 --s 750

a jewelry necklace formed by a perfect combination of arcs, semicircles, and dotted lines, silver, gold, and jade, in a minimalist style --v 5.2 --s 550 --style raw

cuboid elements necklace， different sections of the necklace take on the shape of cuboids, intertwining at various angles,minimalist --v 5.2 --s 550 --style raw

jewelry pendant design, several thick arcs lines surround a large diamond in a rotating shape, minimalist style --v 5.2 --s 550 --style raw

第 10 章 Midjourney 机甲、铠甲与生肖设计 52 例

机甲、铠甲与生肖创作关键词

机甲与铠甲设计常用关键词

armor 装甲、weapons 武器、hydraulics 液压装置、joints 关节、cockpit 驾驶舱、engine 引擎、thrusters 推进器、power source 能源、mechanical 机械的、prototype 原型、pilot 飞行员、hud 头盔显示屏、AI (artificial intelligence) 人工智能、exoskeleton 外骨骼、sensors 传感器、targeting 瞄准、defenses 防御系统、mech 机甲、metallic 金属质感、cyberpunk 赛博朋克、military 军事风格、biomechanical 生物机械、sci-fi (science fiction) 科幻、futuristic 未来感、holographic 全息投影的、androide 机器人、sleek 流线型、intriguing 引人入胜的、technologies 技术、future 未来感、rivets 铆钉、wear and tear 磨损、cracks 裂纹、paintwork 涂装、pipelines 管线、led lights led 灯、头盔 helmet、面甲 faceplate、护颈 neckguard、肩甲 pauldron、护胸甲 breastplate、护腹甲 fauld、护臂甲 vambrace、护手甲 gauntlet、腿甲 greaves、靴甲 sabaton、盾牌 shield、骑士 knight、铠甲 armor。

机甲设计著作艺术家

hideo kojima：著名游戏制作人，其“合金装备”（metal gear）是机甲经典之一。

yoji shinkawa：创作了“合金装备”漫画作品。

sunrise studios：创作了“高达”（gundam）等许多机甲主题的动画作品。

masamune shirow：创作了“攻壳机动队”（ghost in the shell）。

neill blomkamp：尼尔·布洛姆坎普导演了“机器人总动员”（chappie）。

shoji kawamori：shoji kawamori 创作了机甲动画“超时空要塞”（macross）。

james cameron：詹姆斯·卡梅隆执导了机甲主题电影“外星人 2”和“阿凡达”。

the wachowskis：安迪与拉娜·沃卓斯基导演“黑客帝国”（the matrix）。

josé padilha：导演了“机械战警”（robocop）。

guillermo del toro：吉尔莫·德尔·托罗导演了电影“环太平洋”（pacific rim）。

除以上艺术家外，还可以参考研究以下艺术家的作品。

syd mead、h.r. giger、simon bisley、ian mcque、nivanh chanthara、craig sellars、rado javor、josan gonzalez、karl kopinski、paul richards、dave rapoza、donato giancola。

十二生肖关键词

鼠 rat、牛 ox、虎 tiger、兔 rabbit、龙 dragon、蛇 snake、马 horse、羊 goat、猴 monkey、公鸡 rooster、狗 dog、猪 pig。

需要注意的是，“鼠”和“牛”在英文中通常被翻译为 rat 和 ox，而不是 mouse 和 cow，在但描述时，也可以使用 mouse 和 cow。

机甲与铠甲设计 35 例

<https://s.mj.run/s7qdnducpnm> full body portrair:: military female mech, liquid metal, vibrant metallic finish, sitting on bench, photorealistic --ar 11:16 --s 200

<https://s.mj.run/le2kznc_q3a> full body image:: super ultraman mech, multi colored camouflage, liquid metal, vibrant metallic finish, cyberpunk background, standing on podium, photorealistic --s 250 --ar 11:16

<https://s.mj.run/gyzvqsbvzhs> mech skull only, multi colored materials, liquid metal, vibrant metallic finish, on table, cyberpunk backdrop, very detailed, photorealistic --s 550 --ar 1:1

digital art of a magestic fierce sci fi robot tiger, ultra realistic, 8k

<https://s.mj.run/yrmqqnwtuyc> full body portrait:: appleseed ex machina briareos, humanoid mech, multi colored materials, liquid metal, vibrant metallic finish, standing in street, cyberpunk backdrop, very detailed, photorealistic --s 450 --ar 11:16

<https://s.mj.run/t41iq_163eq> full body portrait:: ten story tall super heavy mech, multi colored materials, liquid metal, vibrant metallic finish, standing in street, cyberpunk backdrop, very detailed, photorealistic --s 350 --ar 11:16

<https://s.mj.run/gyzvqsbvzhs> mech skull only, multi colored materials, liquid metal, vibrant metallic finish, on table, cyberpunk backdrop, very detailed, photorealistic --s 550 --ar 1:1

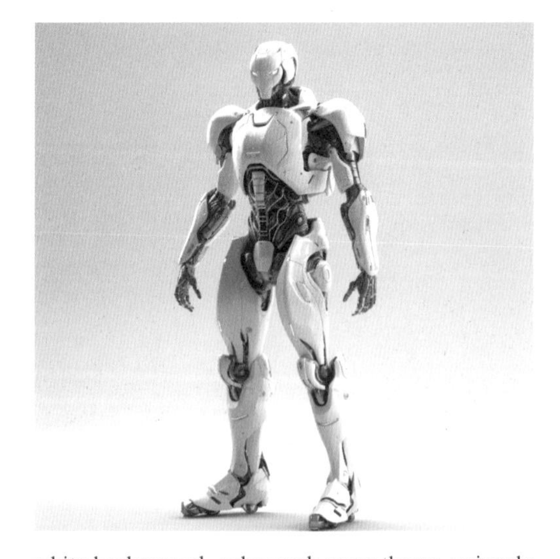

white background, cyberpunk game theme, uniquely designed robot character, 3d full body, clear image, detailed design, 8k, hd, hdr

front view translucent cybernetic tron robot humanoid, georgia o'keeffe, klimt, lichtenstein, --ar 2:3 --s 1000

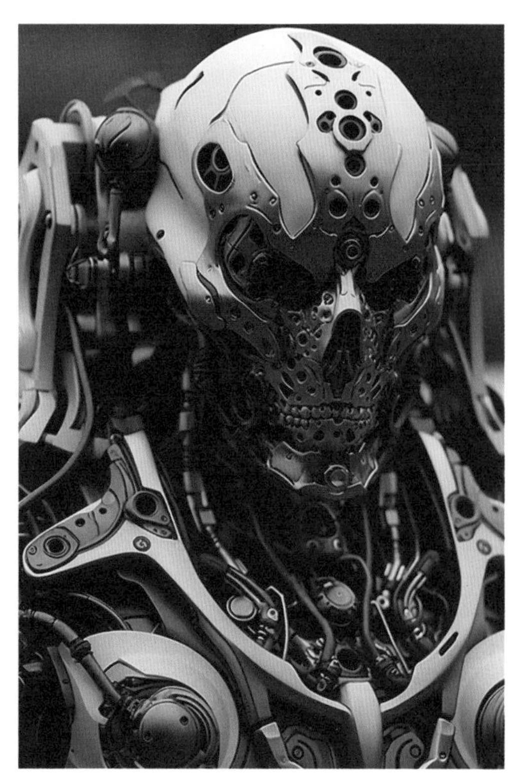

https://s.mj.run/z6h3kot037g mech head only, multi colored materials, liquid metal, vibrant metallic finish, on table, cyberpunk backdrop, very detailed, photorealistic --s 550 --ar 2:3

<https://s.mj.run/hlpfegpjgd0> full body portrait:: robot mech assassin, capable of hovering in mid air, in the style of military equipment, multi colored materials, liquid metal, vibrant metallic finish, standing in street, cyberpunk backdrop, very detailed, photorealistic --s 200 --ar 11:16

hyper detailed clear auto bot transformer in a tech facility talking to scientists on elevated platform, auto bot talks to military scientists at nasa , in the style of mechanical vehicle , --ar 2:3 --s 500 --v 5.2

minimalist luxury female robot concept , hi - tech synthetic bio ethereal metal and black titanium material,super smooth shell,front view --ar 2:3 --s 500 --v 5.2

futuristic, steampunk highly detailed, warframe, red and blue transformers, interstellar mechanic

terrifying scifi futuristic cyberpunk ancient knight robot. platinum colors, photographystudio lighting --s 750

transparent glass, girl android, mechanical, futuristic, nebula, ergonomic, iridescent, holographic, futuristic, --ar 2:3 --s 500 --v 5.1

ai presenting optimized empathetic communication system, very detailed --ar 2:3 --upbeta --q 0.5

a girl standing in front of asilver gray futuristic robot,full body，hyperrealism, frontal, mecha, girl, wearing white and silver gray biker suit one-piece, white martin boots,

model face, cute, kpop girl, cyber knight, high detailed, still from film, digital machine. --s 750 --q 2

sleek future maid robot, designed by ai. female. close-up. intriguing 3d elements. studio lighting. sharp reflections. blurred background objects. ultra photo realistic. . --ar 2:3 --s 1000 --v 5.1

epic action scene featuring a group of futuristic cyborgs engaged in a fierce battle, with intricate details on their armor, weapons, and other cybernetic enhancements. the composition should be dynamic and chaotic --ar 2:3 --s 1000 --v 5.1

ai technologies future

artificial intelligence in humanoid head with neural network thinks. ai with digital brain .technology background concept.female

enchanting cyborg goddess adorned with ethereal dripping paint and blooming flowers, split screen mixed media marvel, pantone watercolor and digital collage, semi-realistic anime style, dynamic angle, soft skin with translucent wires, dripping white milk --s 500 --v 5.2 --ar 2:3

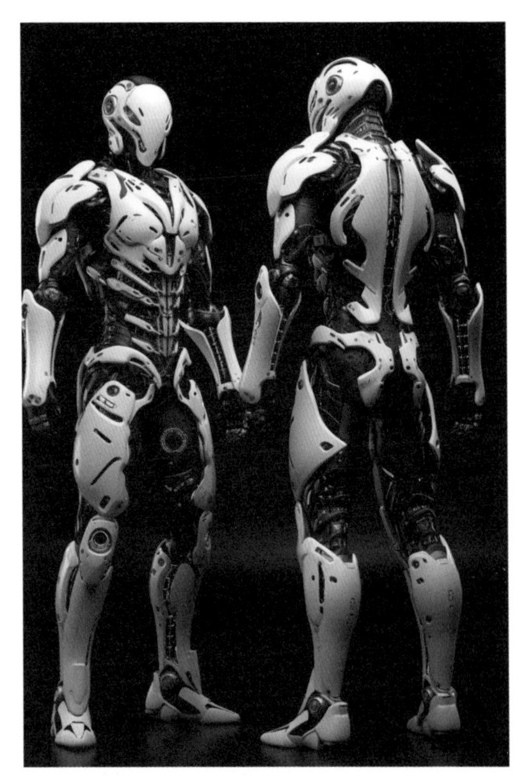

two looks of a robot in cyberpunk armor. front side, back side, an alien warrior in a futuristic outfit holding a gun, in the style of light white and dark silver, on the back are separated shoulder blades, machine-like precision, 3d robot model, full-body, cyberpunk-style armor design, more mechanical details, --q 2 --ar 2:3 --s 750

sorayama style painting of a beautiful robot with shiny reflective metallic skin --s 750 --v 5.1

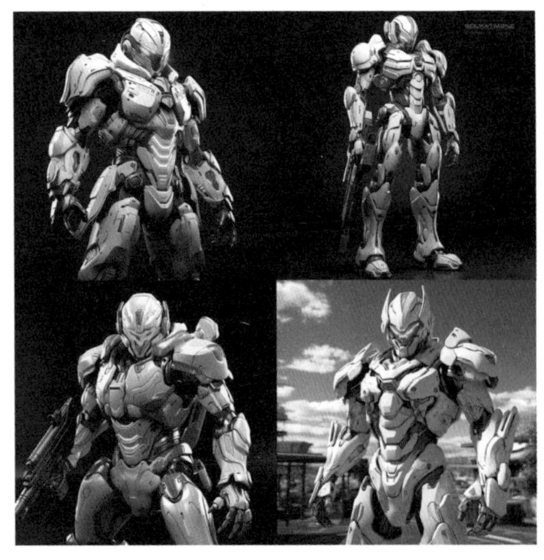

character showcase page, android hunter, dynamic weapon pose, in the style of phantasy star online

full body, futuristic knight in shining armor standing in ruins, h.r giger style, intricate designs etched into the armor, gold and silver accents --s 800 --v 5.2

warrior woman with metallic dragon armor, both the pauldron and the breastplate have dragon head，pauldron is a dragon head --v 5.2 --s 750

model face, cute, kpop girl, cyber knight, high detailed, still from film, digital machine. --s 750 --q 2

<https://s.mj.run/gy2p_1ep72y> <https://s.mj.run/mq_kgye8kpa> beautiful stunning cyborg with cybernetic headress and armor::10 details and composition by greg simkins, claude monet, salvador dali::5 golden ratio, --ar 2:3 --stylize 1000 --iw 1.5 --q 2 --v 5.1

angel in aztecn style armor, white and blue and gold, wings,gold rivets, high detail,flame pattern --s 500 --v 5.1

a beautiful fantasy warrior elaborate armor with many gemstones and jewels --s 500 --v 5.2

sailor moon art perfect face girl on the moon, in the style of cyberpunk futurism, mecha,light violet and gold, bold, heavy shining metal,vibrant colors, close up,hyper photorealistic --ar 2:3 --v 5.1 --s 1000 --q 5

chinese young girl androids, 3d style face, fair skin, exquisite facial features, hanfu and mecha, projection holographic display, surrounded by holographic ribbons throughout the waist, chinese martial arts movements, holographic halo, fluorescent, ,cyberpunk, --ar 2:3 --v 5.2 --s 300

生肖设计 17 例

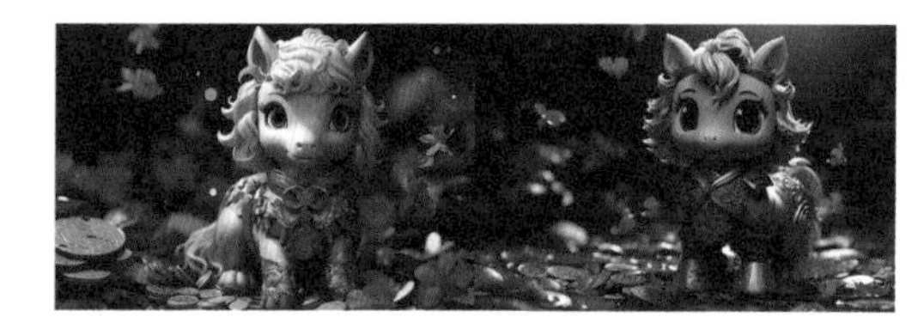

cute little horse dressed in elegant red hanfu, holding gold coin, the ground covered with of gold coins, full body

tiger made of black copper metallic combined with black wood material delicately blend together, full body,wall decor,light gray background

an orange mouse combined with glittering rhinestones, full body, light gray background

pig green gold jade material , full body, light gray background

a fiery monkey , in dark silver and dark orange, cybermysticpunk, dark gray background

portrait of an antique statue of a tiger , golden elements, by j.c. leyendecker full body, light gray background --s 750 --v 5.2 --ar 3:2

3d of a curly hair tiger in a blue jacket, headset and jeans, full body --s 750 --v 5.2 --ar 3:2

3d horse , in the style of surreal cyberpunk iconography, viktor vasnetsov, mechanical designs, by michael whelan full body, light gray background

a white monkey with glasses reading a laptop in bed, by gustave doré, by dr. seuss light gray background

green pig in elaborate fruit arrangement style, full body, light gray background

a tiger with red fur and gold coins, in retro reproduction style, concept art, fancy dress, full body, light gray background

tiger by ivansn, in the style of light orange and light gold, luminous 3d objects, fantastic creatures, translucent color full body, light gray background

the gold buffalo, crafted from gold using ancient wire - carrying techniques, full body, light gray background

tiger a lucite crystal ball and golden stars, light gray background

buffalo in the style of detailed fantasy art, white and crimson, by amir zand, gemstone art , by kerem beyit , full body, light gray background

buffalo made of ebony and yellow rose wood,in style of aztec art, by mike mignola, swirling shining vortexes line on body, full body, light gray background

tiger in style of samurai ,made of gold and silver and copper full body, light gray background

第 11 章 Midjourney 特效文字设计效果 48 例

特效文字设计应用与制作思路

特效文字素材可以应用于海报设计、广告设计、UI 设计、LOGO 设计等各个领域。在制作方面主要是依靠改变文字的材质、造型。

制作特效文字除了可以使用 MJ 外，还可以使用由 Adobe 推出的 Firefly，网址为 https://www.adobe.com/sensei/generative-ai/firefly.html。

除此之外，也可以 Stable Diffusion WEB UI 配合 controlNet 插件制作字效，关于这种技术的文章，可以在网上轻松找到。

特效文字设计 48 例

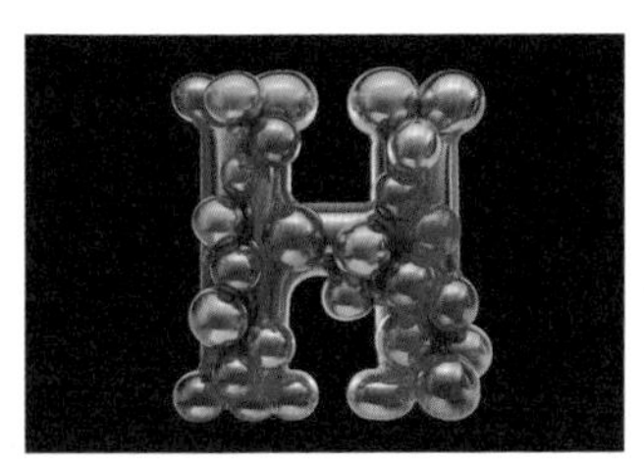

letter h composed of transparent and colorful hydrogen balloons --v 5.1 --s 750 --ar 3:2

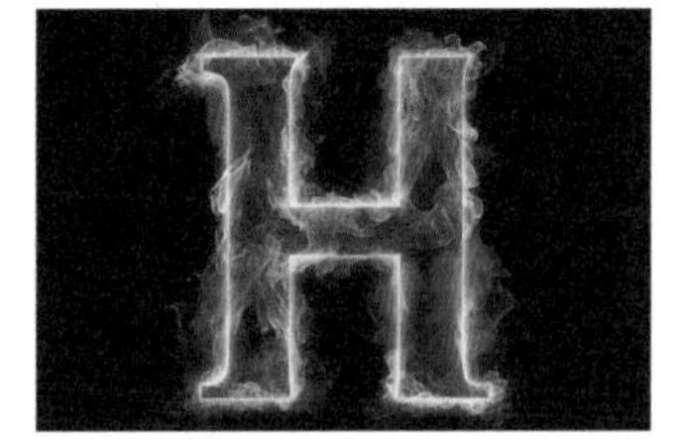

letter h made of red and orange flaming lightning --v 5.1 --s 750 --ar 3:2

letter h in style of emblem ,eagle, vector,fashion,in 2300s --v 5.0 --s 750 --ar 3:2

letter h covered with butterflies, hyper-realistic details --v 5.1 --s 750 --ar 3:2

letter h made of cracked concrete, exaggerated expression --v 5.1 --s 750 --ar 3:2

a futuristic version of the letter h --v 5.1 --s 750 --ar 3:2

letter h in a glass sphere, neon,led lighting --v 5.1 --s 750 --ar 3:2

letter h made of blue mosaic and gold, cubo-futurism --v 5.1 --s 550 --ar 3:2

letter h made of yarn handmade lace weaving , wool material --v 5.1 --s 750 --ar 3:2

the letter k , white background . futuristic hacker style --v 4

the letter x, fancy, ornate, white background, 4d, --v 4

the letter k, steampunk style, 4k, no shadow, white background --v 4

the letter k, multi dimensional paper cut craft, paper quilling, colorful flowers, ornate --v 4

the lettre s , borderlands, cell shading , cartoon --s 750

roman warrior, the lettre s, esport, white background --v 4

the letter s in a bold and heavy serif typeface with beautiful ornate flourishes and embellishments

the letter s chinese dragons , white background --v 4

the letter x made of crazy fire and ice , white background --v 4

3d letter m made by water --ar 3:2

number 2 made by tree roots and leaves isolated --ar 3:2

advertising art of the number 5 composed of leaves --ar 3:2

number 6 made by a beautiful flowers, celebrate 4 years of marriage, white background, make space between the numbers --ar 3:2

3d number 6, clean clear glass, monochromatic, white background, cinema4d, octane render, clear edge definition, prismatic forms

3d letter "h" in style of beautiful baroque --v 5 --s 350 --ar 3:2

night,3, number symbol 3 , bold, 3 is the smooth thickness of the glass,transparent, pink and white and blue holographic, --v 5.1 --s 750

3d neon text style number 3, pink and purple colour theme, glowing, dramatic, powerful, clean

gold numbers "5" with white, pink, and dark pink flowers and pink peonies

two kitchen knife that make the letter "x" for logo branding, vector art, illustration

letter "r",3d modeling, colorful transparent glass, caustics,optics,inflatable,cyberpunk lighting, black background --v 5.2 --s 750 --ar 3:2

a letter "r" made of white balloones, black background --v 5.2 --s 750 --ar 3:2

a photo of the letter, c with fruits and vegetables packed inside of it

back to school teacher design letter a letter b

cute cartoon letter a as a single letter, lisa frank inspired, make it glow, add sprinkles, --niji 5

coloring page for children, alphabet letter uppercase a, no shading, no details

create letter m in the styel of paper origami, --v 5.2

sign with the letters 'og' using masonic symbols

desing a business logo incorporating the letter 'b' inside a jigsaw puzzle. have the puzzle slightly broken with ome pieces yet to be placed. used pastel colours

a bar logo in the style of a van gogh painting with bottles and the letter f on a black background

a minimal letter "p" logo with 3 minimal 2d rounded building blocks squares stacked on eachother in the shape of a step. make background purple. 2d. plain color

create the letter h out of different materials --v 5.2 --s 750

golden letter w. in style of ancient aztec temple aesthetics --v 5.2 --s 750

letters h in single plastic block on a white background --v 5.2 --s 750

letter "p" in pixar movie style --v 5.2 --s 750

the letter a in corroded metal and red paint spatter, polaroids, letter a, flash photography, horror, terror, surreal --s 250 --v 5.1

<https://s.mj.run/wco86dxwuko> in a casual game style a red metal letter a

no color, no shading black and white capital letter d with dinosaurs around it for coloring books thin crisp black lines

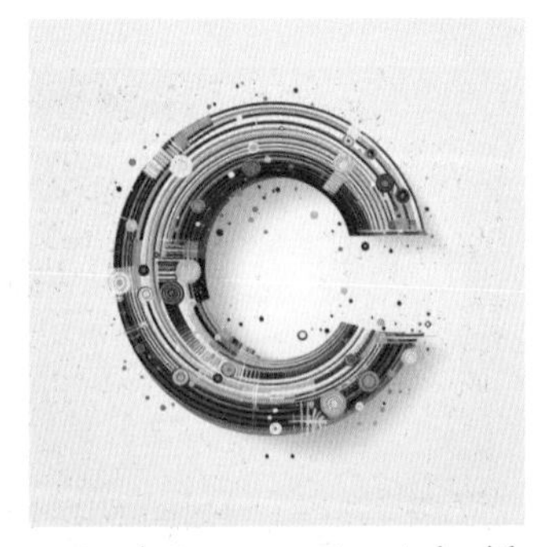
vector letter c constructed with circular lines that propagate in continuous lines

lego letters e on white background

第 12 章 Midjourney 室内设计 26 例

家居室内设计常用关键词

不同房间关键词

卧室 bedroom 客厅 living room、厨房 kitchen、浴室 bathroom、餐厅 dining room、书房 / 办公室 study/office、洗衣间 laundry room、车库 garage、阁楼 attic、地下室 basement 、走廊 hallway、壁橱 closet 、食品储藏室 pantry、游戏室 playroom、阳光房 sunroom、客房 guest room、婴儿房 nursery、健身房 exercise room、影音室 / 家庭影院 media room/home theater。

室内设计常用关键词

室内设计 interior design、空间规划 space planning、家具 furniture、照明 lighting、色彩搭配 color palette、纹理 texture、地板 flooring、墙面材料 wall coverings、装饰元素 decorative elements、布局 layout、材料 materials、织物 fabric、室内软装 upholstery、美学 aesthetics、比例 proportion、功能性 functionality、人体工学 ergonomics、平衡 balance、和谐 harmony、对比 contrast、节奏 rhythm、重点 emphasis、比例 proportion、比例尺寸 scale、统一性 unity、焦点 focal point、对称 symmetry、不对称 asymmetry、重复 repetition、运动感 movement、负空间 negative space、正空间 positive space、过渡 transition、深度 depth、层次结构 hierarchy。

室内设计物件常用关键词

沙发 sofa、咖啡桌 coffee table、餐桌 dining table、椅子 chairs、床 bed、床头柜 nightstand、衣柜 wardrobe、梳妆台 dresser、书架 bookshelf、写字台 desk、脚凳 ottoman、长凳 bench、边柜 sideboard、电视柜 tv stand、镜子 mirror、地毯 rug、窗帘 curtains、照明灯具 lighting fixtures、艺术品 artwork、花瓶 vases、靠垫 cushions、毛毯 throws、时钟 clock、植物 plants、装饰物件 decorative objects、吊灯 pendant light、吊扇 ceiling fan、壁灯 wall sconce、落地灯 floor lamp、台灯 table lamp、凳子 stool、扶手椅 armchair、小型双人沙发 loveseat、控制台桌 console table、陈列柜 display cabinet、脚踏凳 footstool、衣帽架 coat rack、壁炉 fireplace、酒架 wine rack、杂志架 magazine rack、墙艺 wall art、雕塑 sculpture。

常见设计风格提示关键词

中国风 chinese style、日本风 japanese style 、印度风 indian style 、波斯艺术 persian art、古埃及艺术 ancient egyptian art、古希腊风 ancient greek style 、古埃及风 ancient egyptian style 、古代印加文化艺术 ancient inca cultural art、地中海复兴风格 mediterranean revival style、巴洛克风格 baroque style、现代哥特式风格 modern gothic style、土耳其风格 turkish style、简约主义风格 minimalism style、地中海风格 mediterranean style、维多利亚风格 victorian style、高科技风格 high-tech style、罗马风格 roman style、意大利风格 italian style。

家居室内设计 26 例

卧室及浴室设计

interior design for a modern stylish bedroom, leaf design on bedding, lime green and bright green design

neoclassical bedroom, black brutal aesthetics interior for men, with classic art, white peony flowers, a lot of books, rich classical interior

russian classical interior design, bedroom--ar 16:9

modern bathroom, lights track, in the style of jeppe hein, curved mirrors, neo-concrete, light black, by hans hinterreiter, by tondo, by kitty lange kielland --ar 3:2 --s 450 --v 5.2

a typical spanish luxurious cabin bathroom designed by josep puig i cadafalch constructed and white and gold aesthetic, --ar 3:2 --s 450 --v 5.2

客厅设计

living room ultra modern design with yellow and blue color scheme ,a modern interior gray walls concrete luxurious penthouse , architectural ,wood design furniture --ar 16:9 --q 2 --s 500 --v 5

living room, designer edgy brooklyn apartment, muted,modernist furniture , moldings around door and windings and celling edges --ar 16:9 --q 2 --s 500 --v 5

new modern chinese room, natural light wooden elements neutral tones,plants , local handicrafts warm textures textured rugs ,elegant chinese lamp --ar 16:9 --s 500 --v 5.2

modern interior in style hi tech interior --s 750

living room in black with a grey couch, in the style of intentionally canvas,dark gray, eye-catching

儿童房设计

interior design for a kids room in the style of nasa --ar 3:2 --s 250

realistic photo of kids blue orange room with posters on wall --ar 3:2 --v 4

kids room, sports them,with slides, climbing walls --ar 16:9 --s 500 --v 5.2

little girl's room design, a large area of white, a small area of pink, cute princess style theme, cute bear graphics printed on the bed sheet --ar 16:9 --s 500 --v 5.2

沙发、椅子设计

the sofa chair is painted in bold primary colors, with a square geometric pattern, leather, fine texture, and avant-garde shape --v 5.2 --s 750

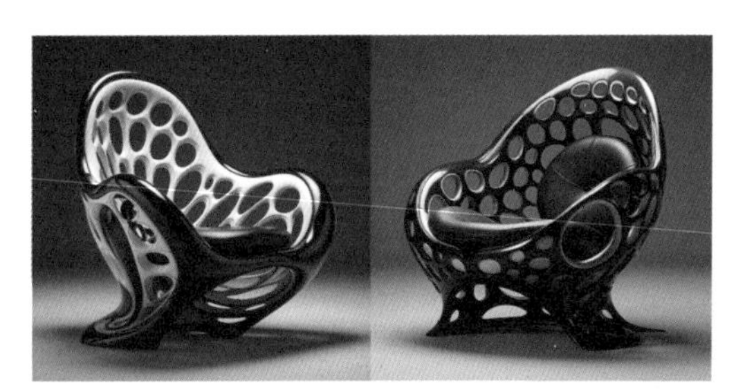

armchair, intricate black organic shape, openworked, mesh structure, surface decorated with white arcs --v 5.2 --s 750

futuristic gaming chair with led and computer screen --ar 3:2 --s 250 --v 5.2

a round sofa inspired by the wireframe sofa by industrial facility for herman miller,more geometric form. the thick cushions should all be blocky --ar 3:2 --s 200 --v 5.2

chair all made by one piece of curved suface, one or two hollowed part on the chair, simple but innovative shape, super clear, studio lighting, 16k, white background

an armchair in the style of jelly, made of jelly, pink color --s 250

花瓶、书架、灯具设计

design a flower stand made of wood, inspired by the incense burner ding, combined with the function of the flower stand. the base of the trellis can be designed with a circular base of the incense burner ding. the stand of the flower stand can present curves and carved details --v 5.2 --s 750 --ar 3:2

wooden vases with green plant , classic japanese simplicity, translucent planes, limited color range, arched doorways, hard-edge style, light black and brown --v 5.2 --s 750 --ar 3:2

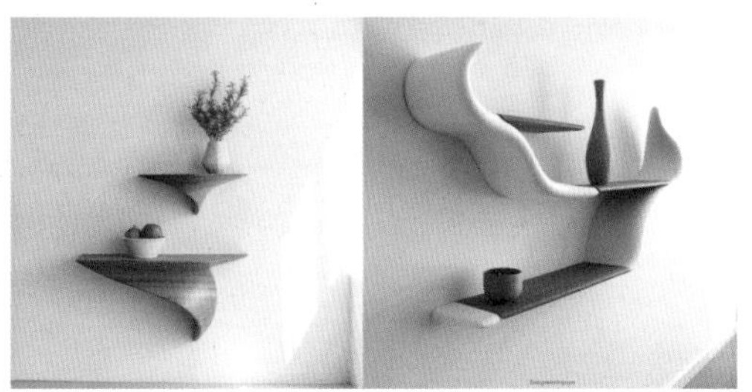

white wall with a wood shelf in an interesting shape that is not a regular flat shape

tetra design library shelf, irregular curvilinear forms --ar 87:70 --s 750 --v 5.2

a lamp with a wavy shape on top of a white table,elegant and feminine gesture in the style of wood sculptures, captivating lighting, --v 5.2 --s 750

a designer bedside lamp in bauhaus style, placed on a desk, creates warm and cozy indoor lighting with depth of field and blurred background against a modern and minimalist backdrop, 8k --q 2

第 13 章 Midjourney 游乐场、卖场、建筑外观创意 27 例

建筑设计常用关键词

常见建筑类型关键词

房屋 house、公寓楼 apartment building、别墅 villa、商店 shop、餐馆 restaurant、办公楼 office building、学校 school、医院 hospital、图书馆 library、超市 supermarket、酒店 hotel、教堂 church、寺庙 temple、银行 bank、城堡 castle、剧院 theater、博物馆 museum、体育馆 stadium、电影院 cinema、地铁站 subway station、机场 airport、城市公园 city park、购物中心 shopping mall、游泳池 swimming pool、桥梁 bridge。

常见建筑构件提示关键词

墙壁 wall、地板 floor、屋顶 roof、窗户 window、门 door、楼梯 stairs、柱子 column、梁 beam、框架 frame、地基 foundation、瓷砖 tile、地毯 carpet、护栏 railing、扶手 handrail、天花板 ceiling、地台 terrace、橱柜 cabinet、吊顶 suspended ceiling、隔墙 partition、阳台 balcony、壁炉 fireplace、管道 pipe、窗帘 curtain、吊顶 suspended ceiling。

常见的材料提示关键词

铝合金板 aluminum alloy panel、铜板 copper plate、不锈钢 stainless steel、玻璃 glass、彩色玻璃 stained glass、陶瓷 ceramic、人造石材 artificial stone、金属板 metal panel、石材 stone、砖 brick、木材 wood、哑光不锈钢 matte stainless steel、大理石 marble、铝板 aluminum panel、钢 glass、金属板材 metal panel、陶瓷 ceramic、石材 stone、橡胶 rubber、金属网 metal mesh、太阳能板 solar panel、镀锌钢板 galvanized steel plate、玻璃幕墙 glass curtain wall、陶瓷板 ceramic plate、石材 stone、瓦片 tiles、人造石材 artificial stone、玻璃钢 fiberglass、铝塑板 aluminum-plastic panel。

知名建筑设计师提示关键词

弗兰克·盖里 frank gehry、扎哈·哈迪德 zaha hadid、福斯特爵士 norman foster、雷姆·库哈斯 rem koolhaas、让·努维尔 jean nouvel、比亚克·英格尔斯 bjarke ingels、伦佐·皮亚诺 renzo piano、圣地亚哥·卡拉特拉瓦 santiago calatrava、贝聿铭 i. m. pei、赫尔佐格和德·梅隆 herzog & de meuron、安藤忠雄 tadao ando、丹尼尔·利伯斯金德 daniel libeskind、史蒂文·霍尔 steven holl、彼得·祖姆索尔 peter zumthor、理查德·迈耶 richard meier、卡洛·斯卡尔帕 carlo scarpa、勒·柯布西耶 le corbusier、阿尔瓦·阿尔托 alvar aalto、路易斯·康 louis kahn、米斯·凡·德·罗 hemies van der rohe、隈研吾 kengo kuma 。

游乐场、卖场、建筑外观创意 27 例

游乐场设计

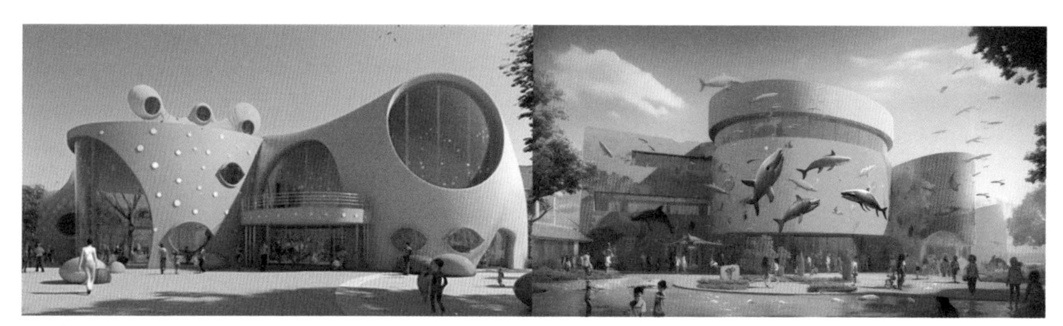

architectural design,the theme of children's entertainment games, colorful,design with marine creatures as the main theme, such as dolphins, whales, penguins. the color scheme mainly consists of blue and green, in a wildflower meadow. designed by rem koolhaas and oma --ar 16:9 --q 2 --s 850 --v 5

architectural design,the theme of children's entertainment games, the theme of star wars, is located in the park, colorful, aerial view. --ar 16:9 --q 2 --s 850 --v 5

卖场空间设计

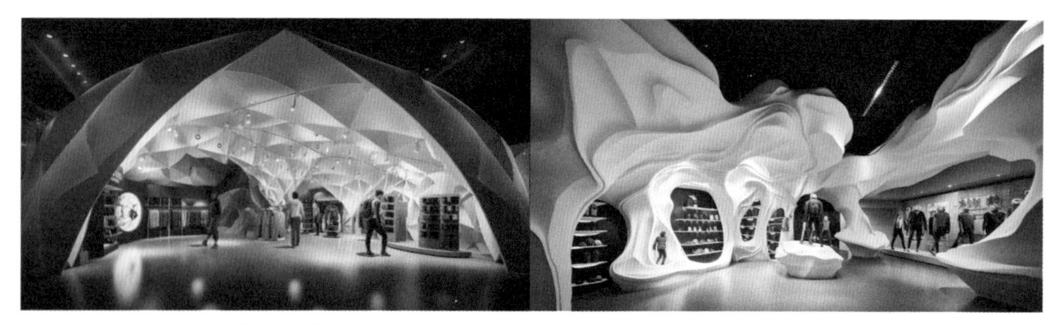

sports gear superstore with rock climbing wall, nike store, kayaking, camping gear, rustic, warm environment, modern architecture, futuristic, zaha hadid design --v 5 --ar 16:9 --q 5

immersive retail space designed by unstudio ramp with areas of third spaces

photograph of the front view of a pathway inside a shopping mall leading into the interior of a supermarket area. --s 750 --v 5.2

餐厅空间设计

a futuristic underwater restaurant concept at the bottom of a cruise ship. large glass windows offer views of the ocean from beneath the waves

restaurant design, with multiple white led lights on the roof, stylish white seats, a small rectangular fish pond in the middle of the restaurant floor, and aztec patterns on the walls of the restaurant --style raw --s 750 --v 5.2

summer terrace on the roof of a one-story house dug into the mountain. forest around, rest area, terrace, bar, lighting, restaurant on the roof --s 750

mountain restaurant inspired by shanty island by paul moenning, renzo julia eddie fabriker i, in the style of vray tracing, ethereal nature scenes, dark, foreboding landscapes, kintsugi, translucent layers, frontal perspective, --s 750 --v 5.1 --style raw

别墅设计

a two-storey villa taken by a professional photographer, the exterior walls of the villa are painted in off-white color, giving it a clean, elegant look. front view. no text, less plants, no logo --v 5.2

mies van der rohe style contemporary villa, the facade is made of honeycomb steel plate, 3 storeys high, floor-to-ceiling windows, external spiral staircase --s 750 --v 5.2

interior view of villa in bali's forest, contemporary, beton brut and glass, solar screens in staves of alternating white stone slabs, minimalistic, front view, design pool,use the sytle of mathias klotz architect --s 750 --v 5.2

a villa with a mix of mediterranean and european styles, a parking lawn in front of the door, and a virginia creeper on the metal exterior wall. --s 750 --v 5.2

modern wooden house on mountain forest cliffs in chinese style

minimalist chinese-style villa in the clouds for mid-journey, natural light and soft lighting,sharp focus, natural transition, unreal engine --ar 4:5 --s 750 --q 2

公共空间设计

outdoor workspace with shade from the sun in a public crouded street in the city, organic shapes, white and orange color --s 250

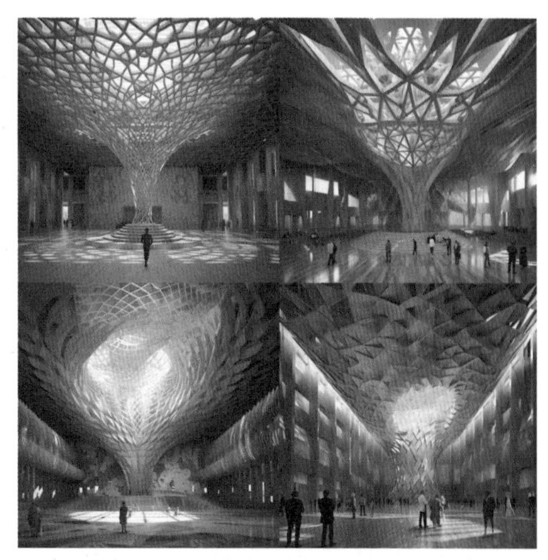

a large modern structure with elaborate ceiling and decorations, in the style of adam martinakis, wood veneer mosaics, shilin huang, multifaceted geometry, talbot hughes, elongated shapes, biblical grandeur --s 750

tensile white fabric structure pavilion, interior, in a park on a summer day, people having picnic, lecture with podium and rows of chairs, zaha hadid --v 5

桥梁造型设计

inhabitable bridges between river, by poalo soleri and santiago calatrava, flowing lines, curved structure --s 250 --ar 16:9 --v 5.2

an artist's rendering of a bridge crossing river, in the style of new british sculpture, eye-catching, spirals and curves, engineering/construction and design, grey concrete infrastructure, oshare kei, pretty --style raw --v 5.2

建筑外观创意设计

photograph by paul zizka and jimmy nelson, architecture concept art, airport in a native american futuristic alien arctic world, architecture by frank gehry, morphosis, aerial view, control tower, terminals, airplanes, tarmac, runway, taxiway --style raw --ar 3:2 --v 5.2

contemporary cultural center featuring dynamic geometric architecture and flexible exhibition spaces, creative architecture --ar 16:9 --style raw --s 250

aerial photography, downtown football stadium, future city, cyberpunk, skyline, long exposure,night --ar 3:2 --style raw --s 750

展厅设计

an exhibition stand for a futuristic copper mining company called dcm. the color scheme should include mostly black accentuated by a few copper colored elements. the stand should include models of giant excavators and mines

a booth for an expo with a modern look, made out of wood and steel structures, very clean design, 2 tv screens on the side, a desk on the right side, people standing infront

前卫建筑概念设计

round cosy futuristic home by ocean during sunset --ar 3:2 --s 250 --v 5.2

supercar exhibition center design, parametric architecture, architectural shape is inspired by ufo --ar 16:9 --s 450 --v 5.2

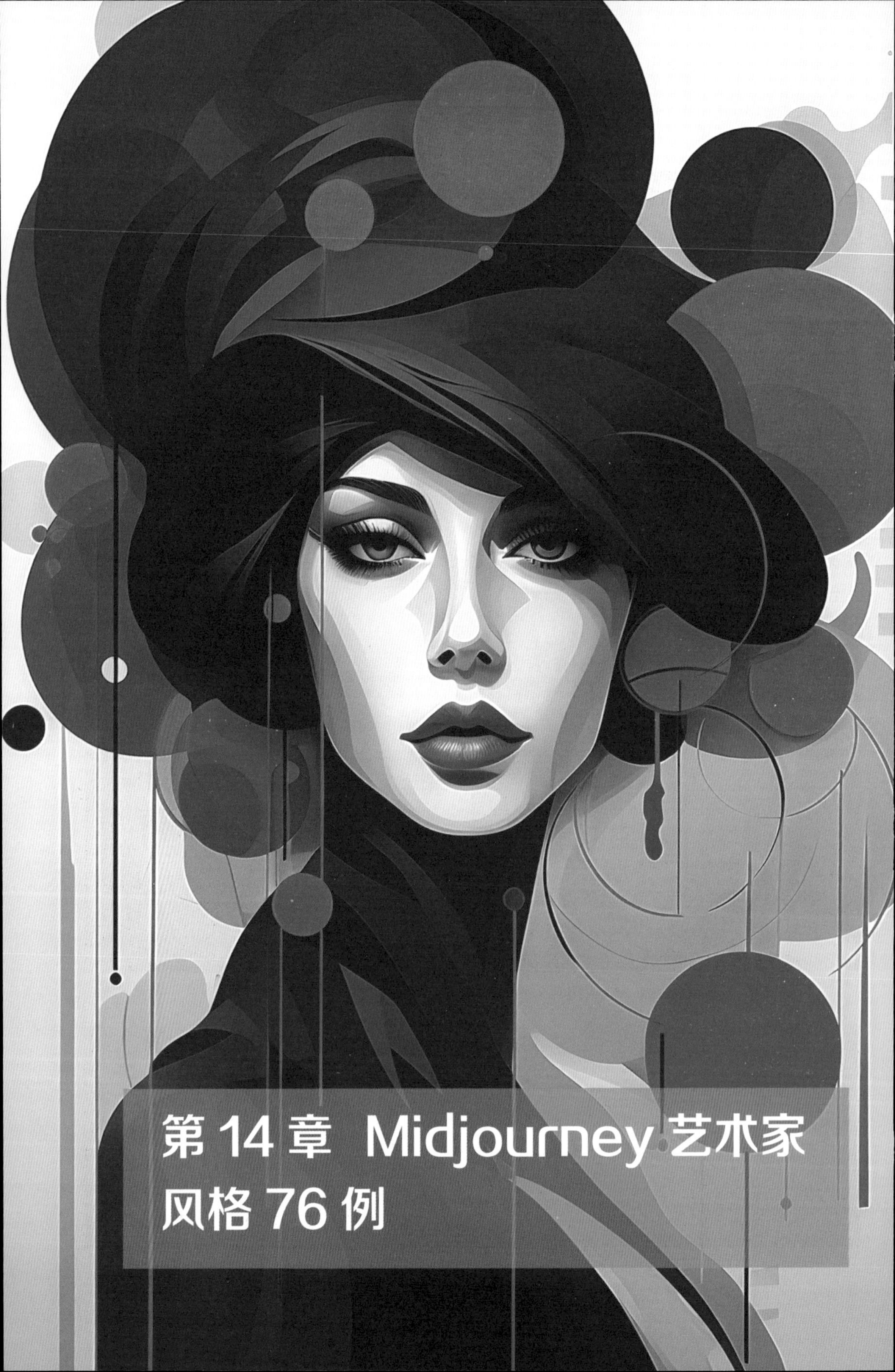

第 14 章 Midjourney 艺术家风格 76 例

了解艺术家关键词对图像的影响

理解艺术家对图像的影响

不同的插画家或设计师有着不同的特色，例如，carne griffiths 的作品通常是用墨水、茶或酒等液体媒介绘制，然后用彩笔或水彩增强色彩和细节。他的作品具有流动性和流畅感，融合自然和抽象元素且有一定的纹理感和层次感。又例如，经典动画《final fantasy》系列的角色设计和插画设计师 yoshitaka amano 的作品则具有细致的线条和纹理、梦幻和浪漫主义及动漫特点。MJ 的 AI 模型经过训练，可以模仿不同艺术家的创作风格，因此，在提示词中加入不同的艺术家名称，能使生成的作品具有明显的风格倾向。

艺术家名称关键词使用注意事项

需要注意的是，MJ 并不能模仿所有艺术家的风格。首先是因为艺术家创作的作品数量极其庞大，受成本约束，MJ 在训练模型时不可能对所有艺术家进行模仿训练。其次由于有些艺术家的创作手法与风格相近，MJ 无法精准地判断两者间的区别，因此只能选择风格变化明显的艺术家来训练。

另外，大量艺术家创作的作品内容比较有局限性，例如，某一些艺术家只绘制植物类作品，因此在描述人物的场景中使用此艺术家的名称，并不会对生成的图像产生明显的影响。

综上两点，在创作中使用艺术家的名称时，要注意使用知名的、风格有特色的艺术家，而且创作的作品，最好与艺术家作品内容类似。

艺术家名字及作品检索网站推荐

https://www.wikiart.org/en/artists-by-art-movement

https://topimage.design/

https://gallerix.asia/

https://www.artic.edu/

https://www.allhistory.com/painting

日式插画常用艺术家及作品关键词

新海诚 makoto shinkai，作品《你的名字 your name》；手冢治虫 osamu tezuka，作品《铁臂阿童木 astro boy》；宫崎骏 hayao miyazaki ，作品《千与千寻 spirited away》；尾田树 eiichiro oda，作品《one piece 海贼王》；武内直子 naoko takeuchi，作品《美少女战士 sailor moon》；井上雄彦 takehiko inoue，作品《灌篮高手 slam dunk》；平井久司 hisashi hirai，作品《高达 gundam》；藤本弘 hiroshi fujimoto，作品《哆啦 a 梦 doraemon》；吉成曜 yon yoshinari，作品《新世纪福音战士 evangelion》；鸟山明 akira toriyama，作品《龙珠 dragon ball》；三浦美纪 miki miura，作品《樱桃小丸子 chibi maruko-chan》。

艺术家风格 76 例

为节省版面，如果图像使用的描述语与前一个图像一样，仅仅是更换艺术家，则其下方将仅有艺术家名称。

male and female model eraserhead art, in the street,by paul klee --v 5.1 --ar 9:16 --s 750

dancer eraserhead art, in the street,by paul klee --v 5.1 --ar 2:3 --s 750

male and female model dreamworks art, in the street,by jasmine becket-griffith --v 5.1 --ar 9:16 --s 750

male and female model dreamworks art, in the street,by ryohei has --v 5.1 --ar 9:16 --s 750

by stephen gammell --v 5.1 --ar 9:16 --s 750

by takato yamamoto

by leon kossoff --v 5.1 --ar 9:16 --s 750

by makoto yukimura

by john frederick kensett --v 5.1 --ar 9:16 --s 750

by hundertwasser

by käthe kollwitz

by goro fujita

by moebius

by odd nerdrum

by paul lehr

by hsiao ron cheng

by frank miller

by dan mumford

by thomas kinkade

by victo ngai

by gustav klimt

by gustav klimt

by agnes bruce pennington

by gediminas pranckevicius

by lisa keene

by joan miró

by carne griffiths

by caspar david friedrich

by hubert robert

by carne griffiths

by andreas rocha

by amedeo modigliani

by chris foss

by marcell chagall

by jason limon

by grzegorz domaradzki

by tomokazu matsuyama

by wassily kandinsky

by edward hopper

by lawren harris

by luis royo

by sailor moon

by hiromu arakawa

by charles blackman

by lisa keene

by bauhaus

male and female model warrior fighting eachother dynapic pose moulin rouge art, in the battle field, by sandra chevrier --v 5.1 --ar 9:16 --s 750

by jules bastien - lepage

by peter bagge

by john berkey

by hope gangloff

by quentin blake

aerial view, in style of eric carle ,illustration,many detail,a pirate stands on a very high hill, looking down at the whole city --ar 9:16 --v 4

in style of tomokazu matsuyama

in style of boris vallejo

in style of rufino tamayo

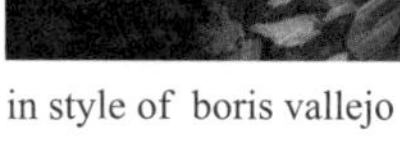

in style of boris vallejo

in style of rufino tamayo

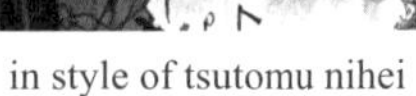

in style of tsutomu nihei

in style of yoshitaka amano

in style of by kawacy

in style of margaret mee

in style of carne griffith

in style of erin hanson

in style of ghibli cartoon style

in the style of killian eng

in style of luis roy

in style of frank frazetta

in style of mike mignola

in style of john singer sargen

a girl is walking on the street.in style of chibi maruko-chan, by miki miura --ar 2:3 --niji 5

.in style of gundam,by hisashi hirai

in style of spirited away ,by hayao miyazaki

in style of your name ,by makoto shinkai

in style of eguchi hisashi

in style of yoji shinkawa

第15章 Midjourney版式、盲盒、包装、UI等设计74例

菜单、价目表版面设计

food menu template design without texts by adobe illutrator --s 1000 --v 5.2 --style raw

coffee shop price list design draft with gorgeous external european border, graphic design work, 2d design, made with illustrator, printable, coffee background color, hand-painted style pattern --ar 16:9 --s 750 --v 5

文创类冰箱贴设计

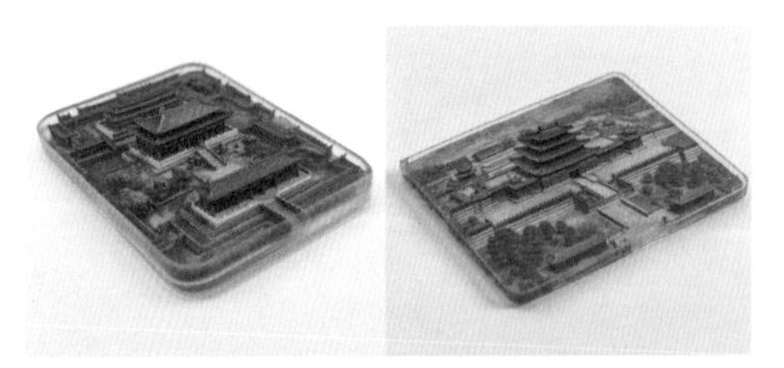

transparent plastic shaped fridge magnet of forbidden city --s 800 --v 5

a glass 3d round fridge magnet with sculpted forbidden city --s 800 --v 5

a 3d round fridge magnet with sculpted forbidden city --s 800 --v 5

a irregular edge 3d fridge magnet of forbidden city --s 800 --v 5

宣传册、杂志版面设计

a magazine layout with dynamic typography and engaging visuals

create a magazine layout for a travel publication, showcasing stunning destinations and captivating stories.

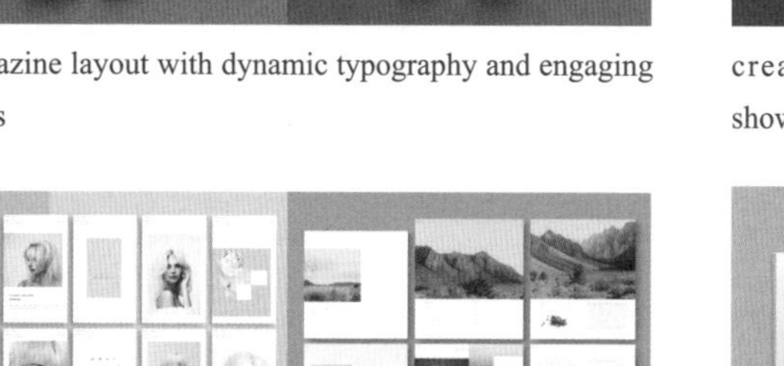

create multiple long pages of squarespace blank template for pinterest blogger, ux, ui, best principles of web design, best principles of ux and ui design

editorial design for a double-page design review magazine, with several sections, with a geometric and minimalist design in orange colors --s 250 --v 5.1

表情包设计

the various expressions of cute cat, emoji pack, multiple poses and expressions, happy, sad, expectant, laughing, disappointed, surprised, pitiful, aggrieved, despised, embarrassed, unhappy, 3d art, c4d, octane render, white background --v 5

the various expressions of cute dragon, emoji pack, multiple poses and expressions, happy, sad, expectant, laughing, disappointed, surprised, pitiful, aggrieved, despised, embarrassed, unhappy, cartoon style, white background --v 5

有文字空间的招贴底图设计

fresh vegetables on black background. variety of raw vegetables. colorful various herbs and spices for cooking on dark background, copy space, banner --ar 3:2 --s 250

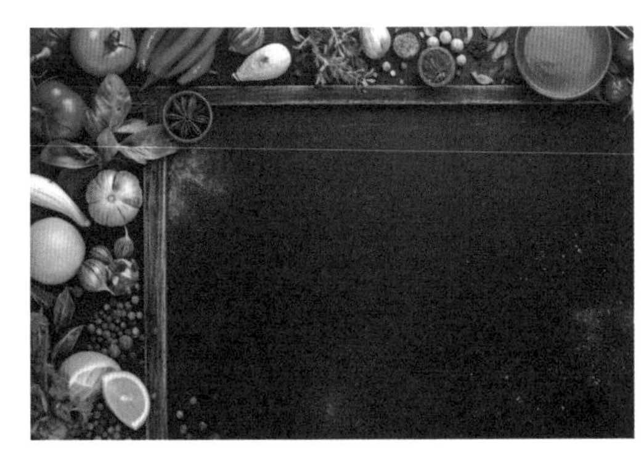

topdown view of a chalkboard placed on a dark wooden table surrounded by healthy food and spices, beautifully distributed, 4k, --ar 3:2

top view of various makeup and cosmetics accessories, pink background, centered, lots of copy space, flat lay, realistic, minimalism, beautiful, modern --ar 3:2 --s 750

four red burning candles with shiny lights at corner, red and gold dot background , copy space --ar 3:2 --s 750 --v 5.1

样机素材设计

mockup image ,blurred beautiful woman pointing finger at a mobile phone with blank white screen --ar 3:2 --v 5

empty picture frame, sitting on wood floor next window , thin black frame mockup

边框素材设计

traditional complex chinese border frame design, vector, pattern, muted tone --v 4 --q 2 --ar 1:2

delicate detailed primrose and forget-me-not flowers border design, copy space, pastel blue and yellow colour theme, victorian style --ar 8:5 --v 4

a gold circle with flower, watercolor ,white background --s 750 --v 5.1 --c 5 --v 5.1 --s 750

a gold circle with long dragon, watercolor ,white background --s 750 --v 5.1 --c 5

剪贴画素材设计

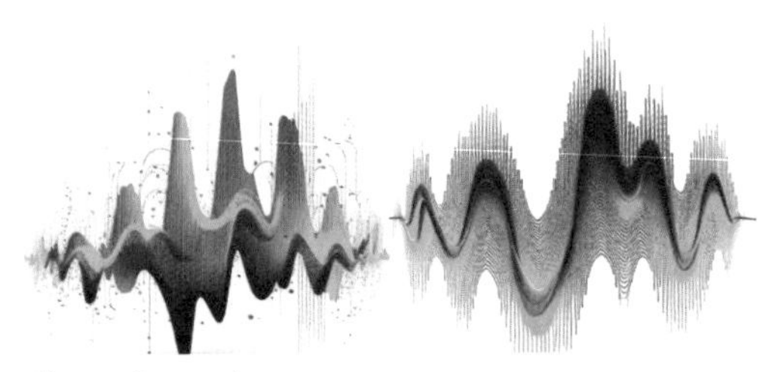

clipart of a sound wave

clipart of suitcase with a pretty sun hat hanging on the side --v 5 --q 2

set of watercolor romantic wedding clipart, white background --ar 9:16 --s 500 --v 5

set of watercolor romantic forest cute little animals clipart, white background --ar 9:16 --s 500

无缝拼贴图案素材设计

flat wallpaper pattern, art deco, fibonacci sequence --tile --v 5

cute cat,flat wallpaper pattern, pink --tile --v 5

黑白矢量效果素材设计

a simple shape based black vector symbol of koi fish isolated on a white background, minimalist design, sleek lines, and sharp edges. using only black and white colors. --v 5 --s 750

a simple shape based black vector symbol of kung fu isolated on a white background, minimalist design, sleek lines, and sharp edges. using only black and white colors. --v 5 --s 750

pine tree on a white background, black and white, silhouette, vector illustration

simple vector, teaching and being taught, monochrome, meeting, business, white background

科技感矢量效果素材设计

a business illustration of deal concept ,blue color, by freepik, vector art, white background, financial success, upward trend, profit, economy, clean lines, dynamic, infographics, --v 5 --s 600 --ar 3:2 --no text --v 5

a illustration of teacher concept ,blue color, by freepik, vector art, white background, teacher and students, clean lines, dynamic, infographics, --v 5 --s 600 --ar 3:2

徽标设计

outdoor badges design monterrey mexico --v 4

an art deco emblem for a pottery, include food and fynbos --s 750 --v 5.0

badge design,luxury, crown shape,polygon shape, overwatch style,sci-fi style,golden and red --v 5

badge design, mothman silhouette in shades of blue and white, set against a circular background with a blue and gold border.

emblem design,eagle, vector,fashion,in 2300s --v 5

badge design, motorcycle club ,speed,side view , illustration, vector,hope,happy --no dark --v 5

LOGO 设计

business and consulting logo, hr, people connected, company, simple, elegant, modern, blue and white colours, --v 4

logo, accelerate, human resources, abstract, modern, simple, playful, luxurious --v 4

商品展示环境设计

green moss, a platform made of wood slices, and greenery. luxurious environment for advertising products, cosmetics, and other things --ar 3:2 --s 250

circle wood podium in tropical forest for product presentation and cream color background --s 250 --style raw --ar 3:2 --v 5.2

minimal 3d stage background product display podium scene with leaf platform and light from windows, white background with flowers, minimal pedestal for beauty, behind the podium,luxury style, --ar 3:2

light beige, granite, 3d stage scene design, empty product display stand, cylinder display stand, small amount of wheat decoration, --s 550 --ar 3:2 --v 5.2

a 3d render of an empty product display podium inspired by nature, with a wooden texture, gentle curves --ar 3:2 --s 750 --v 5.2

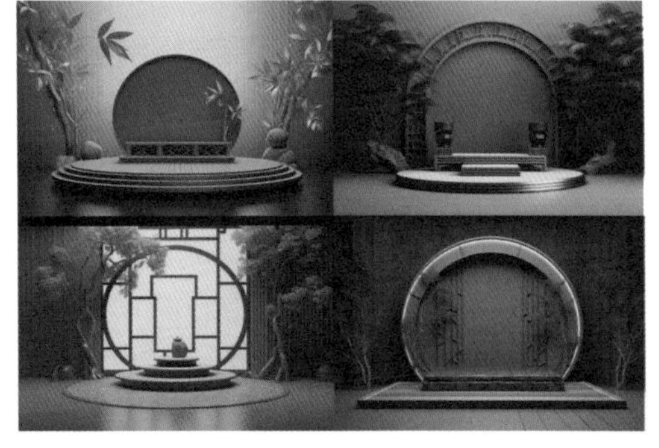

chinese style 3d stage background product display podium scene with leaf platform and light from windows, --ar 3:2 --s 750 --v 5.1

盲盒 IP 形象设计

a blonde winx club girl with pink and blue fairy wings and magnificent fairy outfit, playful pose, kawaii, full body, random neon lights, reflective clothing, clean background, blind box style, popmart, chibi, holographic, prismatic, pvc, --style raw --s 750 --v 5.2

a winx club inspired old mermaid king that looks like triton with a trident, technological sense, playful pose, kawaii, full body, random neon lights, reflective clothing, clean background, blind box style, popmart, chibi, holographic, prismatic, pvc, --style raw --s 750 --v 5.2

a girl wearing little red riding hood on her head, carrying a small basket, wearing a princess dress, black background,playful pose, kawaii, full body, random neon lights, reflective clothing, clean background, blind box style, popmart, chibi, holographic, prismatic, pvc --style raw --s 750 --v 5.2

god of wealth riding a motorcycle, dressed in modern clothing and equipped with mecha-style armor. black background, full body, random neon lights, reflective clothing, clean background, blind box style, popmart, chibi, holographic, prismatic, pvc --style raw --s 750 --v 5.2

地毯图案设计

carpet pattern design,moroccan traditional frame, minimal fashion flower theme --s 500 --v 5.0

carpet pattern design,chinese traditional frame,the theme is chinese mythological story chang'e flying to the moon --s 500 --v 5.0

图标设计

game icon design, 3d design, square shape, with rococo style luxurious metallic rounded corners, semi-transparent glass texture background, a gemstone in center. --v 4

a set of vector graphic of furniture, simple minimal, slack emoji --v 5.1

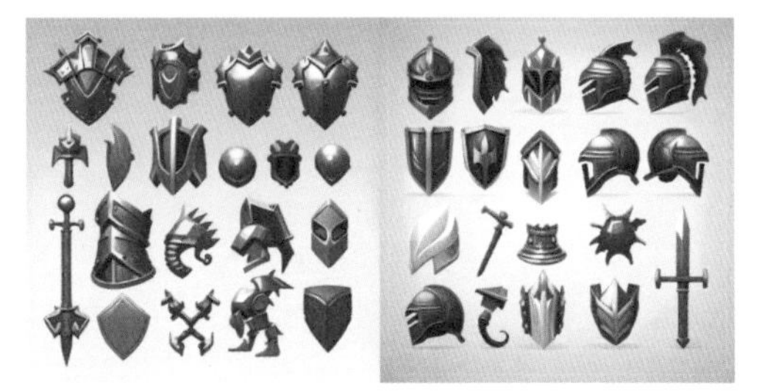

3d,game sheet of different types of medieval armor, white background, shiny, game icon design,style of hearthstone --s 800 --v 4

a 3d glass shield icon, in the style of circular shapes, glass sculpture, isometric, blue and white glaze, high-angle, multilayered compositions, single object, frosted glass texture, glossy base --v 5.2 --s 750

瓷砖纹样设计

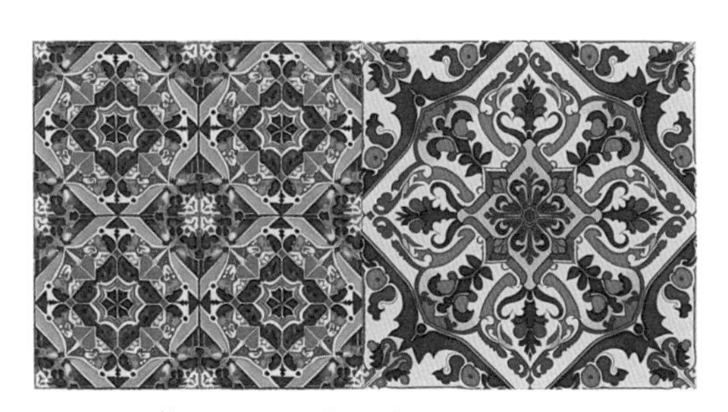

portugese tile pattern --q 2 --v 4

geometrical tile pattern --q 2 --v 4

包装设计

a perfume bottle made of glass withsmooth spiral cone-shaped packaging design sat on a piece of round stone, in the style of flora borsi, golden hues,black background,side lighting --s 750 --v 5.1 --v 5

a perfume bottle made of glass with crescent moon shaped packaging design sat on a piece of round stone, in the style of flora borsi, golden hues,black background,side lighting --s 750 --v 5.1

organic tea brand package design, which mixes gold and silver, uses obvious camellia and tea vector graphics, --s 750 --v 5.2

in the style of product photography, a pump soap bottle is artfully arranged on a bathroom counter, essence of eucalyptus and lavender, farmhouse style architecture, bright and airy natural light --ar 3:2 --v 5.1 --style raw

cosmetic nourishing and hydrating set, natural elements, round label design, baroque design style, gold and gray, product photography, top photography work, --ar 3:2 --s 750 --v 5.1 --style raw

detergent design, product photography, label uses green leaves, sun, garden as main design elements, light color system, minimalist vector design style --s 750 --v 5.2

网页版式设计

personal profile website layout design profesional creative no people vector --s 750 --v 5.2

a webpage for skin care, natural colors and modern design, colours of light sand and shades of blue and green --s 750 --v 5.0

游戏概念场景设计

ruined mysterious alien sci-fi structure, h. r. giger style entrance in the middle of the jungle, walls covered with technology symbol, cosmic sculptures deeply, realistic, highly detailed, 8k --v 4 --ar 3:2

dark fantasy, temple on an ancient mine::5, scattered gold and silver jewels on the ground, detailed, aerial view --ar 3:2 --q 2 --v 4

UI 设计

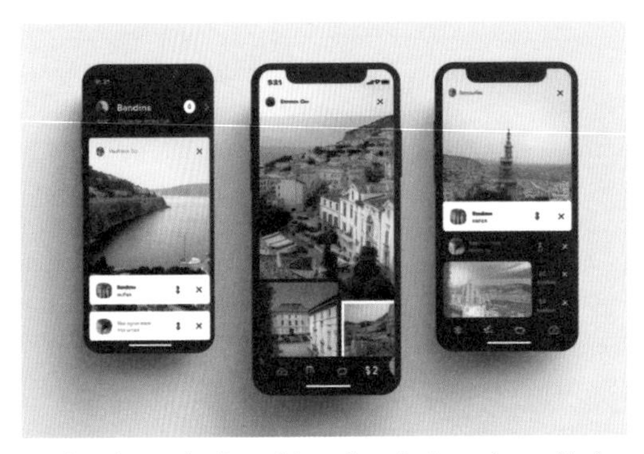

tourist photo viewing ui interface design, photo display main interface, other spaces are comments, photo parameters, front and back page turning buttons --ar 3:2 --v 5.0 --s 750

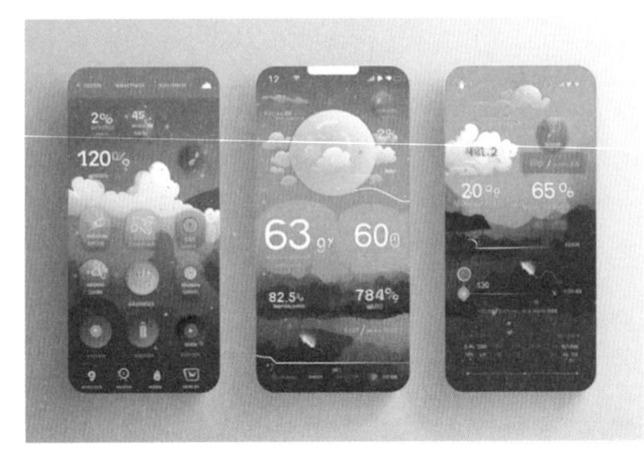

weather view ui interface design, with large icons of different weather, detailed weather condition display board, forward and backward page turning buttons --ar 3:2 --v 5.0 --s 750

卡通头像设计

设计方法是先上传自己的头像，再使用本书第 2 章讲解过的以图生图的方法来制作，例如左图为笔者上传的头像，中间为 3D 效果卡通头像，右图为插画风格的卡通头像。

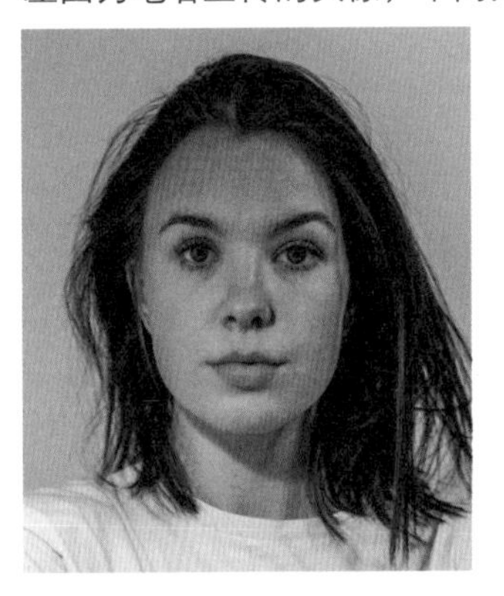

https://s.mj.run/0ifi3zux_ow super cute girl ip in popmart style,white background,3d render --ar 2:3 --v 5

https://s.mj.run/0ifi3zux_ow super cute girl ip design,white background, cartoon style --ar 2:3 --niji 5